AF180717

The Back-Yard Foal

A suckling filly. (Photo by John Boardman)

The Back-Yard Foal

by

Peggy Jett Pittenger

arco
New York

Published by ARCO PUBLISHING COMPANY, Inc.
219 Park Avenue South, New York, N.Y. 10003

Third Printing, 1973

Copyright © 1965 by Arco Publishing Company, Inc.

All rights reserved. No part of this book may be reproduced, by any means, without permission in writing from the publisher, except by a reviewer who wishes to quote brief excerpts in connection with a review in a magazine or newspaper.

Library of Congress Catalog Card Number 72-95504
ISBN 0-668-02748-7

Printed in the United States of America

For

Margaret, Nancy, and David

Preface

ANYONE WHO HAS EVER OWNED A MARE HAS AT ONE TIME OR ANOTHER considered raising a foal or two. In fact it is quite likely that this very possibility determined the selection of a mare, rather than a gelding, in spite of her substantially higher price.

Is it worth it?

Add to the purchase price of the mare the stud fee, board bills, transportation costs, and the charge for a veterinary certificate which is required by the more conscientious breeders. The total may amount to several hundred dollars even before conception occurs, if conception occurs at all, for the domestic horse is notoriously infertile. As a national average there are two mares bred for every live foal produced. The mare must be maintained for almost a year (slightly over eleven months) during her pregnancy; for part of this time her usefulness is diminished. At the very minimum the foal must be cared for another two years before it can do even the very lightest work, or three years from the time that the project was first undertaken. The cost may vary from a few hundred dollars for a farmer who grows all his own feed to several thousand where high board bills are involved. The national average is probably close to $1,500.

The end result may be disappointing if the parents were less than first rate. One moment's carelessness or bad luck can wipe out several years' hard work and planning. Therefore it is usually cheaper and far more practical to buy than to breed, letting someone else put up with the risks, frustrations, and hard work.

In spite of the expense and hazards, it is an incomparable satisfaction to train and ride a horse one has bred and raised. The pleasure increases in direct proportion to the animal's good looks, manners, and ability. Should he actually win a few ribbons or races, the proud owner will be regarded as seer, prophet, and horseman—at any rate by his family and friends.

NOTE

The mares whose pictures are included in this book, along with their foals, are, for the most part, the ones who appeared in *The Back-Yard Horse*. They are:

Blackacre's Deborah (Naja-Dollie D.) 1957, a pinto half-Arab
- Blackacre Grace (Babe) 1963, pinto filly.
- Blackacre Christopher (Devan Marsh Hawk) 1964, palomino colt.

Goldy, circa 1951, chestnut saddlebred.
- Blackacre Sam (Babe) 1962, palomino colt.
- Soleil d'Or (Babe) 1963, palomino colt.
- A Formal Affair (This Evening) 1964, bay filly.

Tippy Dee (Hawk Jim-Tippy Tin) 1944, chestnut Morgan
- Replevin (OCR) 1959, chestnut mare
  - Blackacre Mandamus (Devan Marsh Hawk) 1964, chestnut colt.
- Assumpsit (OCR) 1962, chestnut filly.
- Debt (Devan Marsh Hawk) 1963, chestnut filly.

Beverly Blue (Gay Town-Medina Blue) 1954, bay Thoroughbred.
- Night Season (This Evening) 1962, bay colt.
- Blue Hour (This Evening) 1963, dark bay or brown filly.

Dot Sena (Eternal Road-Irish Channel) 1956, bay Thoroughbred.

Blackacre Beauty (Nipper-Penny) 1959, black Shetland Pony.
- Blackacre Melanie (Little Red) 1964, black filly.

Esterette (Fight A Little-Ethel M.) 1960, chestnut Thoroughbred.

Contents

The Back-Yard Foal

1
Selection of Breeding Stock

ALTHOUGH MANY GRADE HORSES ARE EQUAL OR SUPERIOR TO REGISTERED horses, it is not economically feasible to go into breeding on a large scale with other than registered stock. Even though the initial cost is several times greater, the outlay for feed and equipment is essentially the same. There is a steady demand for good-looking purebred youngsters, while each grade horse must prove his individual merit.

Careful selection of breeding stock is of the greatest importance. Animals should not only be healthy, attractive, vigorous, reasonably young horses, but should also be typical of the breed which they represent. There should be no unsoundness which is attributable to hereditary defects in conformation. Beware of the only good individual from a poor family. The gene pool has often been compared to an hour glass with the top representing ancestors, the narrowest part the individual in question, and the bottom portion the future generations. The less attractive individual from an excellent line with a solid background of desirable traits is more likely to produce acceptable offspring than the only good-looking one from an entire genetic group. Of course the exceptionally handsome horse from an impeccable background is the best risk, but these animals are not often offered for sale at reasonable prices.

Choice of breed is largely a matter of personal taste, as well as the use for which the horse is intended. There is considerable variation in the cost of raising foals of different breeds to a salable age, with the Thoroughbred being by far the most costly. Because they are large, high-strung animals, they burn up more energy and require more feed to maintain condition than cold-blooded horses. The risk of injury is greater because of the racy temperament. Astronomical stud fees, up to $20,000, and the high price of mares with good race records or fashionable pedigrees increases the cost of operation. Thoroughbred yearlings, however, bring correspondingly high prices at auction, the average in 1963 being $5,790. Outstanding colts by leading sires out of well-bred mares which have given a good account of themselves at the track bring correspondingly high prices. But for every Swapson by Swaps out of Obedient, which sold for $130,000 at Saratoga in 1961, or One Bold Bid (Bold Ruler—Forgetmenow), which sold for $170,000 at Keeneland in 1964 for a new world's record, there are hundreds of youngsters on which their owners fail to break even. In areas where fox hunting and polo are not prevalent sports it is hard to dispose of racing culls except at an almost total loss.

Morgans, Quarter Horses, and Arabians are far cheaper to maintain than Thoroughbreds because they are easier keeping. While racing Quarter Horses may command relatively high stud fees, Arabian, Saddlebred, and Morgan stallions rarely stand at a fee over $500. Arabian mares are more expensive than those of any breed whose progeny do not have a chance to earn back their purchase price in racing competition. Morgans are the oldest American breed. They are also the least numerous. Because of their greatly increased popularity in recent years mares of good type are in short supply and great demand, which facts are reflected in the prices asked for them.

In selecting a brood mare prospect look for a full-bodied, easy keeping mare. She should be a good milker; the size and development of the udder is usually, but not always, an indication of the ability to milk. Maiden mares have small udders which do not reach their full development until after the first foal has nursed. Producing brood mares of good type and family are seldom offered for sale, especially if they are in the prime years between five and fourteen. Mares in this age group which are advertised for sale are often culls which the owner is dumping because they have not produced anything noteworthy, because they are barren, hard to settle, or are poor milkers. As a rule it is possible to buy a highly desirable brood mare only at the dispersal of a stud due to death or illness of the owner or to the dissolution of a partnership. Therefore the would-be breeder must settle for a brood mare prospect, a young mare which he hopes will be capable of producing the type of foal he likes. Mares should be inspected by a veterinarian to make certain that they are in good health, free from infection, and have no anatomical condition which would make conception and foaling difficult or impossible.

Although the domestic horse has never been bred for fecundity, fertility definitely runs in families. The daughter of a good producer is more likely

Arabian brood mare.

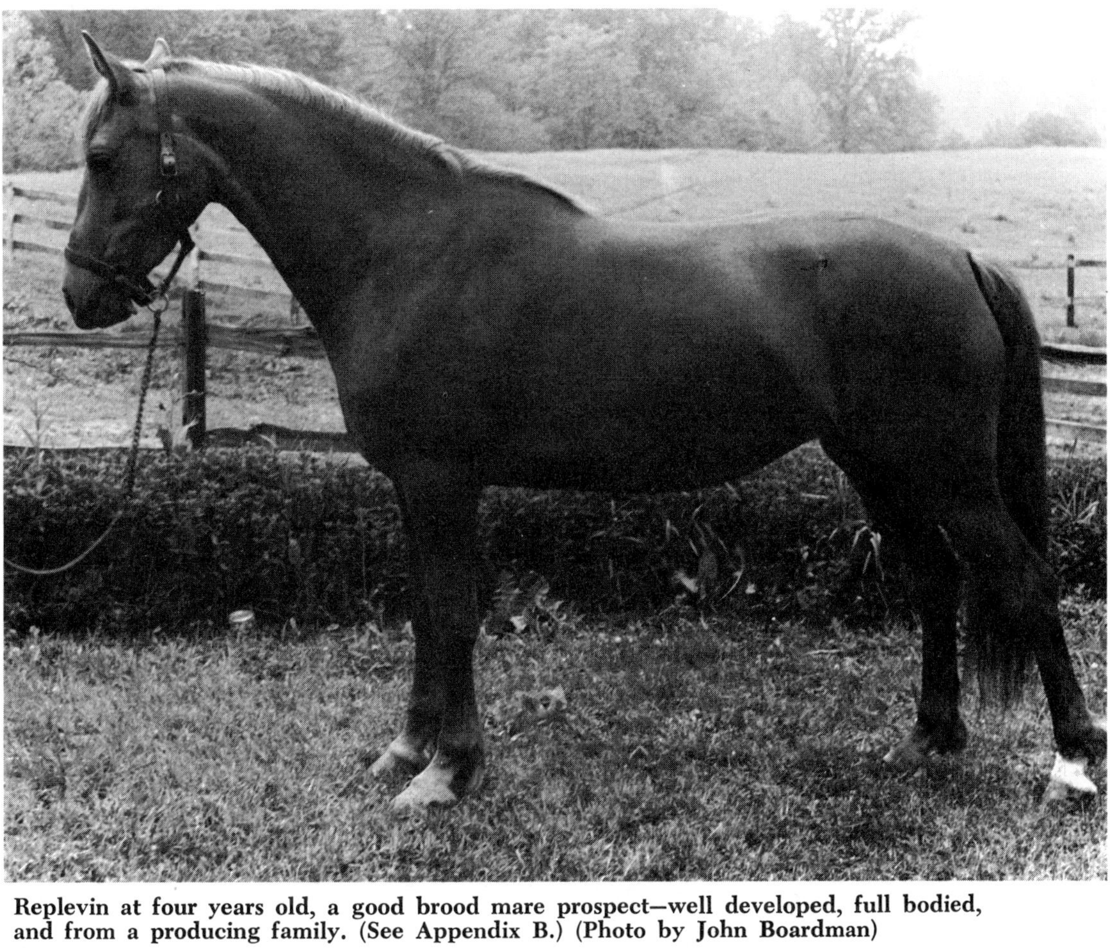

Replevin at four years old, a good brood mare prospect—well developed, full bodied, and from a producing family. (See Appendix B.) (Photo by John Boardman)

to be herself a good producer than is the offspring of a hard-to-settle mare. The dam's produce record should be checked. All other factors being equal, a filly out of a mare which has produced eight or ten foals is a better candidate for the stud than the filly which is the only foal out of a mare that has had ample other opportunities to conceive. This information is available from the breed secretary, from one of several livestock agencies, or may be found in the breed's stud book; there may also be a record of other matings which did not result in a live foal.

The buyer should know that, unless otherwise stated at the time of the sale, the stud fee of an in-foal mare is presumed to have been paid by the seller. All guarantees, warrantees, and return privileges, however, are voided by the sale and should the mare prove to be barren, or should the foal abort or be stillborn, the owner of the mare is entitled to no refund or preferential consideration from the owner of the stallion.

As a mature stallion can cover between twenty-five and forty mares in a season which lasts from February to June, or a hundred mares where breeding is carried on year round, it is considered impractical for the owner of a few mares to maintain a stallion. The sum spent in stud feed does not add up to the cost of buying and keeping a stallion. Some breeders own a stallion jointly. Many stallions are syndicated, with several owners holding one or more shares, each of which entitles him to a breeding season.

It really goes without saying that a stallion must be a healthy, vigorous, typical member of his breed with no hereditary defects. At one time stallions with certain specified unsoundnesses were barred by law from public service. Only the most outstanding representatives of any breed should be kept entire so that the breed will improve rather than retrogress. One of the hazards of a breed's extreme popularity is that males which are not first rate fail to be gelded and are permitted to propagate.

The stallion must be well mannered. Viciousness, for which there is no excuse, is the result of mishandling or a bad temperament which is hereditary. A stallion offered at public service must have a high conception rate or ratio of live foals to mares covered; his get should be of the approved type. A young horse which is not yet proven should have his semen checked for volume, viscosity, and the number and motility of spermatazoa before he enters the stud. The practice of breeding a colt to a few test mares before he officially stands is a time-consuming method, but is the only irrefutable proof that he is capable of siring foals, since some horses with apparently normal sperm are sterile.

Although good brood mares are seldom offered for sale, stallions often are. Many small breeders who maintain only one stallion must change herd sires from time to time as their original horse's daughters come to dominate the brood mare band.

The stallion should be of the same general type as the mare where similar offspring are desired. If the mare has some patent defect it is well to select a stallion which will tend to correct rather than to accentuate it. For instance, a coarse mare bred to a fine stallion may produce a foal of great quality while a mare somewhat deficient in bone would produce a more desirable offspring if mated to a stallion with greater substance. Find a stallion with straight hind legs for the cow-hocked or sickle-hocked mare, but do not expect miracles. Few stallions are prepotent in regard to all characteristics. If a mare has too many obvious shortcomings, she should not be bred at all. Never breed to your neighbor's horse merely because he is available. Be as critical of him as of a horse stabled a hundred miles away. It is a false economy to breed to the local stallion if he doesn't measure up to standard in all the important respects. The extra cost of shipping to a superior horse may prove to be a very worthwhile expense.

It is desirable that a stallion show breed characteristics to a marked degree. On looking at him there should be no doubt in which register he is included. For instance, a Thoroughbred-type Morgan, although a handsome animal in every other respect, would be criticized by Morgan breeders

Hot and cold blood carried to extreme in light horses. The black mare is very coarse with dense hair. The gray gelding is fine-boned, attenuated, and weedy looking with very sparse hair. (Photo by John Boardman)

Thoroughbred-type Morgan. Esterette (Fight-a-Little-Ethel M. by Golden Vein) dam of BURGLARY, multiple stakes winner of over $125,000.

Persistence of Arabian type: dish face in modern Thoroughbred.

Typey Morgan.

as being too weedy, while a Thoroughbred that resembled a Saddle Horse would be condemned for being too high fronted. Not surprisingly, there is a strong resemblance between Quarter Horses and Thoroughbred sprinters. Although earning index generally is considered to be a better guide than conformation in selecting a racing sire where price is no object, many of the best Thoroughbreds are Arab in type: *Khaled, Tetratema, *Mahmoud. Of the last named, winner of the English Derby in record time, sire of stakes winners *Majideh, Billings, Oil Capitol, Mount Marcy, Monsoon, Snow Goose, General Staff, *Flushing II, *The Axe II, etc., the late Aga Kahn wrote:

> I bred him as an ideal Arab horse and would like to add that in breeding Mahmoud I definitely had the Arab ideal before me and for once it came off which was a real triumph when all the millions of orders of hereditary combinations are thought of.[1]

There are two theories in regard to the relative size of the mare and stallion. One point of view considers it unwise to breed a mare to a stallion more than two hands taller than she is, since the large size of the resulting foal would make delivery difficult or impossible for her; too great a disproportion in the size of the genitalia will cause injury to the mare. The other theory holds that small mares may be bred safely by stallions of any size (providing that a breeding roll is interposed to prevent possible vaginal rupture), since the offspring tends to resemble the dam in size at foaling—the exception being the stunted mare of a large breed which is genetically normal size. Certainly many strains of pony and small horse—the Welsh, Connemara, and Criollo, to cite a few—have been vastly improved by allowing Thoroughbred stallions to run with the semi-wild bands of mares, but how many mares were injured or died as a result is not known.

If the stallion stands at any great distance, the mare should be left at his farm for at least three weeks after she is first bred so that she may be tried and bred again if necessary, rather than trucking her back and forth. Not only is the risk of injury in transit lessened, but there is also less likelihood of missing the next heat period entirely if the mare fails to catch on the first service. She should arrive at the stud several days before she is due to come in season. The per diem for the mare will vary from $1.50 to $3.50 depending on the locale and the type of accommodations offered.

There are almost as many different terms as there are stallions standing. Some owners charge a flat fee payable at the time of service or before the

[1] H. H. Aga Kahn, "Foreword," *Thoroughbred Racing Stock*, by Lady Wentworth (London: George Allen & Unwin, Ltd., 1960) p. xvii.

Arabian or Thoroughbred? (Photo by John Boardman)

mare leaves the farm. There may or may not be return privileges within the current season, either free or for an additional fee. Sometimes part of the fee is payable when the mare is bred, the balance when it is determined that the mare has settled, or when the foal stands and nurses unassisted. Some fees are due and payable on October 1st following service in lieu of a veterinary certificate stating that the mare is barren. Some horses stand on a live foal basis, no fee being due until the foal is dropped; or if the fee had been paid previously it is refunded if the foal dies before it nurses; if the owner of the mare prefers, the fee may be applied to a service for the current year. In some agreements the owner of the stallion retains a share in the unborn foal's earnings, or a final portion of the stud fee is payable from the foal's first purse or when he is sold or claimed.

Most stallions stand at half price to grade mares, while some owners will not cover grades with their registered horse for fear that one or two unlikely looking foals will act as adverse advertising and give the stallion a bad name. Some public-spirited owners give service free or at a token fee to 4H club members when the preferred breeding season has passed, or after

June 15. "Private Treaty" may mean whatever the traffic will bear or it may mean free service to a mare with an exceptionally good race or show record in order to build up the reputation of a young, untried stallion by the fine foals he is likely to get on superior mares. Some ads say, "Special consideration to mares with foal at foot," as these mares are obviously sure breeders, obviating the nuisance and profit-shrinking practice of repeated covers for a single fee. For the same reason some busy horses do not stand to maiden mares as it often takes repeated breedings to get them in foal. Most terms state "to approved mares, only," so that the stallion owner may reject any mares of such low quality that their offspring are likely to tarnish the horse's reputation.

When a stallion is sold, his obligations and commitments go with him. Unless otherwise stated, sale does not cancel his future bookings nor the return privileges to mares which he failed to settle or which did not produce a live foal.

2
Care of Breeding Stock

The horse is a product of the grasslands and grass is, therefore, his natural feed.
—EARLE

GOOD PASTURE IS THE BACKBONE OF ANY SUCCESSFUL HORSE-BREEDING PROGRAM. For this reason many of the best horse farms grow none of their own feed, being reluctant to give up any of their acreage to crop production. Not only is grass the most complete ration, it has special functions in addition to supplying energy. Good pasture furnishes vitamins and minerals which may be deficient in cured feeds. The growing green leaves of pasture grasses are rich in protein and also are an excellent source of carotin, which animals metabolize as Vitamin A. Vitamin A deficiency caused by prolonged eating of old or poorly cured hay results in night blindness, rough coat, reproductive failure, difficulty in breathing, and faulty hoof development. Extreme cases end fatally. Many of the common unsoundnesses are thought to be attributable to Vitamin A deficiency.

Horses at grass synthesize Vitamin D by the action of sunlight on their skins, an obvious impossibility for animals which are continually stabled. With the exception of sun-cured hays, most feeds are deficient in Vitamin D. Young, growing stock have the greatest need for Vitamin D to insure the proper development of their bones. Deficiency results in rickets (disordered ossification), in immature horses, or in osteomalacia, characterized by loss of bone density and by bone softness and brittleness in the adult, particularly mares during lactation. Being deprived of sunlight and being fed hay rather than grazing on grass accounts for the lack of substance and high incidence of unsoundness among barn-raised foals.

The horse's mineral requirements must be satisfied. Salt containing

"The horse is a product of the grasslands . . ." (Photo by John Boardman)

iodine and other trace minerals in deficient areas, such as the Great Lakes Region, must be available at all times. It may be dispensed loose in a mineral box, in 50-pound pasture-size blocks or in small 4-pound individual blocks kept in each horse's feed tub or mounted in a wall bracket. Many prepared feeds contain salt. Some old horses with sluggish digestive systems do not drink enough water and extra salt must be sprinkled over their feed to encourage them to drink more and thus prevent impaction. When twenty-year-old Tippy Dee feels the need for additional salt she takes a mouthful of water, moistens her salt block, lets it soak for a few minutes and then licks up the salt which has softened and is now easily available.

Calcium and phosphorus, the demands for which are greatest during the first years of life, during pregnancy, and during lactation, must be supplied not only in adequate amount but in a ratio of twice as much calcium as phosphorus. Deficiencies cause poor skeletal development. An imbalance in the Ca to P ratio with P in excess of 1:2 gives the effect of an actual calcium deficiency. Such an imbalance may be caused by the overfeeding of rich grains and protein supplements. In green forages Ca usually exceeds P making for a desirable balance. Since legumes such as clover and alfalfa are richer in calcium than grasses such as timothy and blue grass, a heavy mixed hay should be fed to balance a heavy grain ration in stabled horses. An unfavorable Ca:P may in part account for the notorious bone softness of sale-prepared yearlings which have been confined and heavily grained.

In general, horses which graze regularly on good pasture have fewer ailments than those which must depend entirely on dry feeds. Even a few weeks at grass has a beneficial effect; it acts as a tonic and freshener—the famous "green grass cure," when stale and overtrained horses are turned out to recuperate.

Pastures should be rotated frequently to prevent their becoming horse-sick; that is, to keep parasite infestation to a minimum and to insure that the grass is of a high quality at all times. Pastures should be mowed twice a year, about June 15 and August 15, to control the growth of weeds, to insure even grazing, and because the fresh young shoots of grass which mowing encourages are richer in vitamins and minerals than longer grass.

Each mare and foal requires two acres of pasture under optimum conditions, though the small pony breeds may get by on considerably less. In the arid regions of the Southwest as many as eighty acres are needed. Pastures must be checked periodically for deficiencies and limed, fertilized, and reseeded as needed. Soil analyses may be obtained from the United States Department of Agriculture Extension Service for a nominal fee.

Overgrazing results in the destruction of grass plants. There is the danger that horses may eat poisonous weeds which they would otherwise scorn.

Fresh water must be continually available.

Brood mares and foals. (Photo by John Boardman)

The effect of overgrazing. Nothing but dust and weeds remains of this once lush pasture.

It is desirable in managing breeding stock that a state of nature be approximated as nearly as possible except for the actual acts of mating and foaling. Certain abnormal cases may need assistance in the form of hormone injections, antibiotics, or even surgery.

Brood mares do best when kept out of doors in spacious pastures the year round. They must have shelter from extremes of weather, but ideally it should be shelter which they can enter or leave at will. Horses suffer far more from wind and damp than from actual cold. Heat and humidity cause great discomfort so that summer shelter is as important as winter. Flies cause constant torment from which a shady place offers some protection; many well-run farms keep mares and foals in their stalls during the day in the summer, turning them out in the cool of the evening when flies are not prevalent.

In the brood mare band there is a distinct social order with one mare being in charge. She walks and the others follow. She drinks first, eats where she pleases and has her choice of shade and shelter. She may kick or bite any member of the herd without risk of retaliation, although the mere laying back of her ears is usually enough to get her what she wishes and to send the others flying. She sometimes asserts her dominance by driving each other mare in turn away from her pile of hay or grain box before she settles down to eat at what she considers to be the best spot. Number 2 mare can bite all the mares except number 1 and has second choice of food and shelter. Number 3 is dominant over all but numbers 1 and 2 and so on down the line to the lowliest "Kigme." The lower on the pecking order a chicken is, the harder he pecks the chickens beneath him;

An ideal paddock for mare and foal: Goldy and her second palomino son.

Thoroughbred mare and foal in a simple run-in shed. This type of shed offers shelter from sun and flies in summer and from snow, wind, and rain in winter. The sand floor provides excellent drainage.

Pastured horses help keep themselves clean by rolling.

Both mare and weanling enjoy being out of doors.

A snow bath. (Photo by John Boardman)

The pecking order: Goldy disciplines the Morgan filly Assumpsit. (Photo by John Boardman)

the same holds true of horses (and humans). The animal at the bottom of the scale, although scorned and sometimes tormented by the animals above it, is often allowed to eat beside number 1, which feels no lessening of status by associating with the proletariat. The most jockeying for position takes place between two rivals for the number 2 spot or when a new mare is introduced and must establish her social position.

At Blackacre, Goldy is the undisputed queen. Any horse turned out with her is deeply marked across the rump by her teeth. Once is usually enough. In any other group Tippy Dee takes charge although she is willing to yield her rule to Goldy without protest. Beverly Blue also heads a group and will yield to Goldy but not as graciously as the old Morgan. It is not safe in this instance, however, to turn the two number 2 mares out together as they are too evenly matched in drive and, although willing to be subserviant to the mighty Goldy, will not yield to each other. A stalemate results—or, rather, perpetual warfare.

A filly usually remains subserviant to her dam even though she has become larger, stronger, and more aggressive with other mares. For instance, Replevin, who is much larger than Tippy Dee, still minds her old mother, although she now bosses Deborah whom she formerly feared. Little Beauty, taken seriously only by the youngest foals, is not considered a challenge by any of the full-size mares; she serves as court jester, eating and standing with impunity wherever she pleases.

Great tact must be taken in introducing a new mare into the brood mare band. There is an inevitable scuffle and shuffle until she finds her level in the pecking order. Usually the new mare must challenge or be challenged by each member of the group to determine which has pecking—or biting—privileges. A timid mare may follow at the end of the line and drink last, but she is less likely to be injured than an aggressive mare that wishes a higher rung on the social ladder. It sometimes works to turn the new mare in a paddock with only the boss hoss until they are fully acquainted. Then when both are returned to the band the new mare will benefit by reflected status.

Mares benefit from daily exercise which they get in a large field in the company of other mares. The mare may be ridden from the time her foal is a few weeks old until a month before the next foal is dropped. Mares which do no work and become too fat are likely to have a more difficult labor than those which are kept fairly fit. In 1962, two Arabian brood mares were entered in the Florida 100-mile Trail Ride. Both placed third in their respective divisions. One mare was eight and a half months in foal; the other was one month in foal and had a two-month-old foal at foot. The first mare was the only horse entered to gain weight, twenty-two pounds, during the course of the ride.

Brood mares, or any other horse for that matter, should never be kept in damp, dark, foul, poorly ventilated stalls. Ammonia fumes from wet, filthy bedding may cause abortion or even death in extreme cases.

Mares vary greatly in their need for concentrates. They will require between one and twelve quarts of oats or the equivalent, depending on size, breed, age, temperament, and condition. For instance, a young, still developing, high-strung, large mare in run-down condition will require many times the grain that a mature, plump, placid pony will need, the pony probably getting by with little or no oats. If oats are scarce and expensive, corn may be substituted for all or part of the ration. Corn supplies twice the energy per measure of oats, so less need be fed, but protein, in which corn is lacking must be supplied—for instance, by feeding legume hay. Many barren mares do well on pasture or hay without the addition of concentrates; it is preferable that they not be allowed to become overly fat. The in-foal mare must be fed enough to meet her own needs and the demands of the growing fetus, which makes its greatest gains in weight during the latter half of pregnancy. She must not put on excess weight, which serves no useful purpose and places an unnecessary burden on heart, lungs, and kidneys. Just prior to foaling the mare should be fed a laxative diet with bran and linseed replacing the other concentrates. Mares with a foal at foot have their grain ration increased gradually, maintaining a fine balance between allowing the mare to go back in condition and causing the foal to scour from excess milk production.

It goes without saying that all feed should be of the best quality obtainable. Spoiled feed will cause indigestion; an estimated 12 per cent of all equine deaths are attributable to colic. If the fungus ergot which grows on spoiled rye grass is present in the hay it will cause powerful uterine contractions resulting in abortion.

A mineral or vitamin supplement is indicated if the quality of the feed is uncertain or if the mare and foal do not seem to be doing well.

In recent years pelletized feeds have been gaining in favor, especially in areas where space restrictions or arid conditions make maintaining adequate pasture unfeasible. In caring for a large number of horses there is considerable saving in time and labor costs by feeding pellets from an Apex self-dispenser. Pelletized feeds, which are marketed under several trade names, contain a variety of ingredients, such as linseed, soybean meal, kelp, brewer's yeast, corn, oats, safflower meal, beet pulp, salt, bone meal, molassas, phenothiazine, ground limestone, vitamins, dried skim milk and sun-cured alfalfa. They may be fed as a supplement to hay and grain to stabled horses or as free choice to pastured horses. Because they are less bulky, pellitized feeds require less storage space than conventional feeds. Although quite a bit more expensive than hay and grain, waste is virtually eliminated. It is maddening to see horses sorting through $50-a-ton hay to find the choicest morsels, soiling and trampeling as much as 30 per cent. Pellets may very well prove to be cheaper in the long run.

Split-rail and board fencing are the safest type of barrier for enclosing horses. When scrap lumber is used, the cost is not appreciably greater than woven wire. Although white paint is decorative when freshly applied, it is expensive; it looks shabby as the paint fades, flakes, and becomes soiled. Care must be taken that the paint contains no lead, for horses chew fences and may sicken or die from lead poisoning. Black tar and creosote paint preserves the wood, needs repainting only every eight to ten years, retains its attractive appearance and costs only 50¢ per gallon at the factory. Barbed wire is not suitable fencing for horses and is especially dangerous for young stock.

In enclosing a pasture for mares and sucklings care must be taken to place the bottom rail or plank close to the ground; otherwise the foal is likely to lie down by the fence and get up on the opposite side, much to

Apex dispenser for pelletized feed. Brood mares at Laguna Seca Ranch, Monterey, California. (Courtesy Apex Company)

the frantic dismay of all concerned. This happens far more frequently than one would imagine.

Shortly before foaling the brood mare is brought up to the barn, where she may be closely observed. She is allowed to exercise in a small paddock where she may be seen at all times. At night she is kept in a spacious stall, preferably at least 14 by 14 feet; the stall must be of generous dimensions so that the foaling assistant and veterinarian, if necessary, will have room to work and maneuver. The stall should be well lighted as most foalings take place at night; there should be some way of heating it for winter foalings in severe climates. Large farms have several stalls especially set aside for foaling which are fully equipped for both normal delivery and emergencies. A cement floor makes easier the thorough cleaning and disinfecting which is necessary before each new occupant. After the floor is disinfected, it is sprinkled with lime and heavily bedded with fresh straw.

The stallion requires a stout stall built of two-inch-thick oak lumber which will not break or splinter should he try to kick his way out. Many modern stallion stables are constructed of cement block, which tends to make damp housing if not well ventilated. The stall should be 14 by 14 feet unless the horse is an unusually large Thoroughbred, hunter, or draft breed, in which case it should be 16 by 16. The stallion stalls of central Kentucky are constructed with palatial dimensions which it is not practical for the small breeder to attempt to duplicate. Ideally, the stallion's stall should have a door opening into a corral to which he has access at all times. All doors must be wide enough, four to five feet, and high enough, nine to ten feet, to prevent injury to the horse or to his handler should the horse lunge, crowd, or rear.

Although a small corral is adequate for exercise, a two- to four-acre paddock is preferable for optimum health and reproductive vigor. Double fences must separate stallion paddocks from other enclosures containing horses, especially other adult male entires. Stallions benefit by regular exercise under saddle or in harness, especially those which show lack of vigor or which practice stable vices. Stallions are fed according to their individual needs. They should not become overly fat.

Breeding stock of all ages require periodic attention to their feet. This is of the greatest importance with foals, weanlings, and yearlings, as their bones are still malleable and may be deformed by irregular hoof growth. Similarly, certain defects in leg angle may be improved or corrected at an early age by judicious trimming and shaping of the feet. Stabled horses must have their feet picked out daily to prevent thrush.

Older horses should have their teeth checked twice a year and floated if necessary. If there is any unexplained loss of condition, the teeth should be examined for irregularities or for decayed (rare) or boken teeth which may interfere with proper mastication of feed.

Older horses are wormed spring and fall for strongyles (blood worms) and in the early winter for bots. Many farms feed low level phenothiazine year-round, rather than worming periodically in therapeutic doses. Young horses require more frequent worming, especially for round worms, which, as a rule, do not infest older horses severely. Horses returning from the track, from show, or from any other situation where they are constantly stabled may become heavily infested with worms in a brief time when first turned out to pasture, as they have not had the opportunity to develop the resistance to parasites which most older horses have acquired.

Parasite control is one of the most important phases of stud farm management—a phase which is too frequently overlooked or neglected by the backyard breeder. Good feed, expensive plant, and blue-blooded stock come to nothing if growing animals are allowed to remain infested with worms which sap their strength and stunt their growth. The principal source of infection in the foal is the mare. The grass in fields and paddocks is continually being contaminated as is hay which is fed on the ground. A rough-

A stallion paddock at Blackacre. Note the five-foot - four panel fence with cap boards.

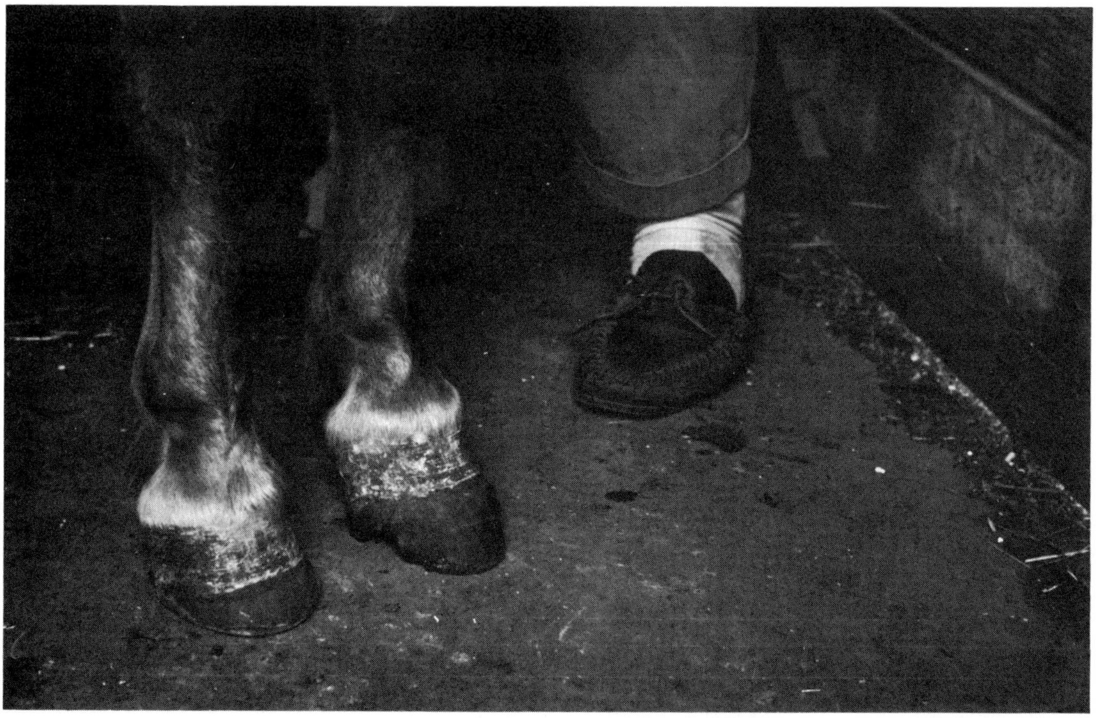

Trimmed hoof, left, and neglected hoof, right.

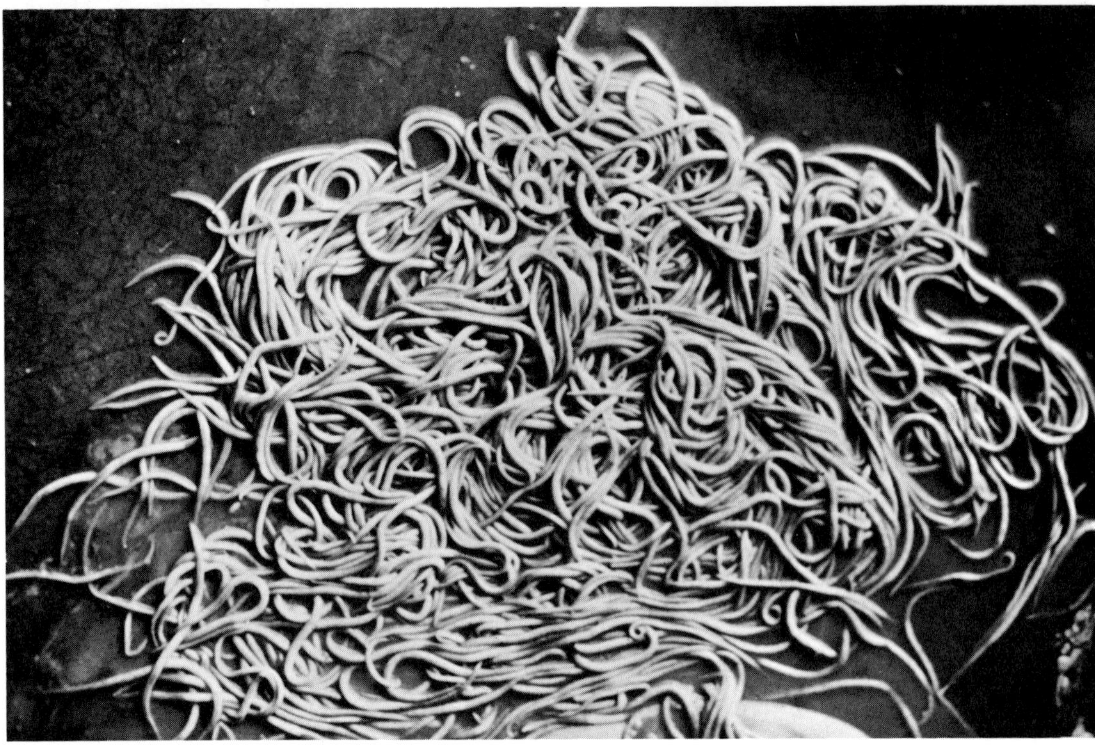

Ascarids (round worms). These were all removed from a single horse. (Courtesy of Fort Dodge Laboratories, Fort Dodge, Iowa)

coated, listless, unthrifty, potbellied look is suggestive of parasite infestation in young horses. Failure to put on weight in spite of an adequate ration is indicative in older animals. The most common internal parasites are:

Round worms (ascarids) do the greatest early damage to foals. They are principally a parasite of young animals, mature horses developing a resistance to them; the exception is the adult horse which has been continually stabled.

The adult round worm lays about 200,000 eggs per day, which pass out with the manure of the host animal. These eggs are very resistant to changes in the weather, surviving extremes of heat and cold. The infective eggs are eaten as the horse grazes on contaminated pasture. In the horse's alimentary tract the eggs hatch into larvae which migrate through the liver and lungs, travel up the trachea and back down the esophagus to the gut. A harsh, dry cough is symptomatic of migrating ascarids. In severe cases of infestation lung damage and pneumonia may result. Large masses of adult worms may cause impaction, colic, or perforation of the small intestines, with subsequent peritonitis and death. Round worms weaken the host animal, making him more susceptible to bacterial infection.

At eight weeks the foal should be treated with piperazine, one ounce per 200 pounds of body weight. The dose should be repeated at regular intervals throughout the first year. Older horses do not require dosing for round worms unless a microscopic examination of a fecal sample indicates their presence.

Blood worms (strongyles, nematodes, red worms) are a group of parasites which is the most dangerous to horses. The larvae migrate through the arteries—hence the name blood worms—causing the vessel walls to thicken. Verminous aneurisms may form in untreated cases, the aneurisms sometimes being invaded by actinobacillus. Adult worms live mainly in the cecum, where they erode the mucous membrane. Eggs pass out into the pasture with the manure. The eggs hatch first-stage larvae which subsequently develop into second- and then third-stage or infective larvae in a period of ten days in warm, damp weather. After being ingested with contaminated grass or hay, the larvae slough off their outer skin, penetrate the gut wall and migrate through the mesentaric arteries before returning to the intestines as adults. Strongyles may cause death by the formation of thrombi, emboli, or aneurisms. Sudden, unexplained death in young horses may be due to coronary thrombosis of parasitic origin or to the rupturing of a verminous aneurism. Anemia and rapid loss of condition are characteristic of blood worm infestation. Strongyles cause colic and impaction by interfering with the blood supply to the musculature of the gut, thereby greatly lessening its tone and efficiency. Foals are treated at twenty-five weeks of age or at weaning with phenothiazine, one ounce per 400 pounds of body weight divided into three daily doses. Adult horses are treated twice a year. Phenothiazine in therapeutic doses is contraindicated for mares during the final months of pregnancy, although they may be safely treated on a low level basis.

Bots (gastrophilis intestinalis) deposit their eggs on the hairs of the legs, where they may be seen as small yellow dots. As the horse scratches his legs with teeth and lips he swallows the eggs, which develop into larvae in the stomach where they form large clusters attached to the mucosa. In large numbers bots interfere with digestion, causing colic and loss of condition. The larvae pass out with the manure, enter the ground and emerge in late summer as adult flies. Dosing with carbon disulfide or with Dyrex removes the larvae from the stomach. To remove the eggs, wash the horse's legs with hot water when the temperature is below freezing for twenty-four hours; the heat causes the eggs to hatch and then freeze in the cold air.

Good management helps control parasites although it is doubtful that they can ever be eliminated entirely. Droppings must be removed promptly

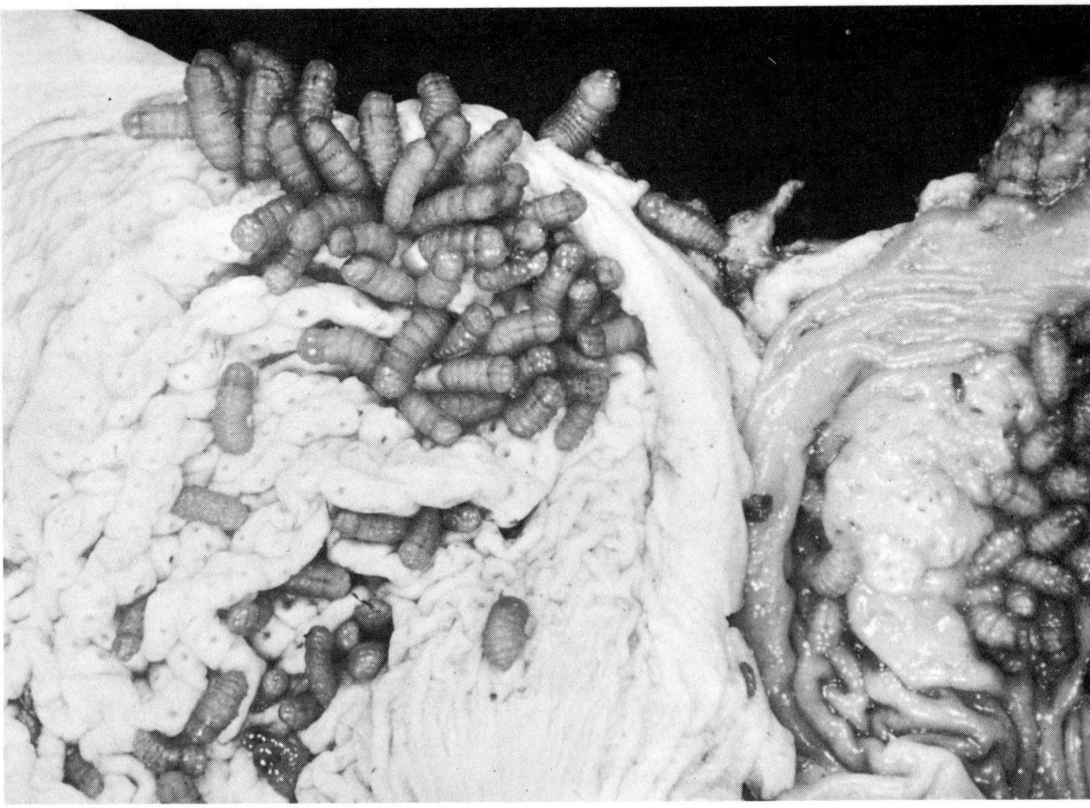

Bots attached to lining of stomach. (Courtesy of Fort Dodge Laboratories, Fort Dodge, Iowa)

from stalls, paddocks, and other small enclosures where otherwise the horse would have to stand more or less continually in his own filth. In large pastures horses practice parasite control by grazing in one area and dropping their manure in another. The practice of scattering manure with a broom rake while encouraging over-all grazing also encourages the ingestion of infective eggs and larvae. Horse manure should not be used to fertilize pasture intended for horses unless it has been stored until it has fermented, the heat thus generated having destroyed the infective material.

External parasites—house flies, stable flies, face flies, deer flies, mosquitoes, lice, mites, etc.—are extremely troublesome. They cause loss of condition both by sucking blood and by tormenting the horses so that they literally have no rest. Certain flies are thought to be the vector in several infectious diseases.

Most flies congregate in sunny areas. A high-potency residual insecticide may be used on their favorite resting places. Horses should be provided with a cool, shady place to stand in hot weather. Manure piles and damp, poorly drained spots are breeding places for flies and should be eliminated. Manure should be kept in a covered pit or bin or else it should be removed from the stable area daily by cart or manure spreader.

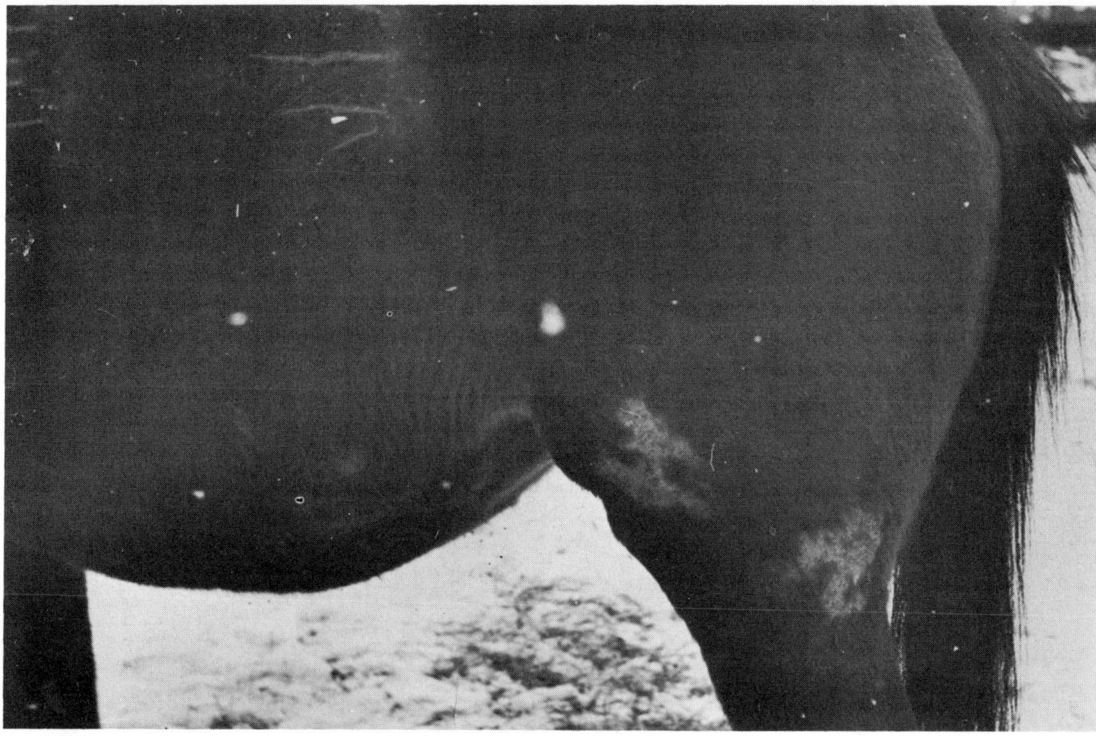

Mange.

Horses may be protected against flies by a spray or sponge especially prepared for horses. Some compounds which are safe for cattle cause blistering and loss of hair when applied to horses.

Lice and mites are most likely to infest dirty, debilitated horses. They are more common in winter than in summer. Zealous grooming, a general improvement in condition, and long hours in the summer sun are often all that is neeeded to get rid of lice. Insecticide powders and sprays are usually effective in more stubborn cases. Sulfur compounds are useful in cases of mange, which is a dermatitis caused by mites.

3
Breeding

THE AVERAGE MARE COMES IN SEASON EVERY TWENTY-ONE DAYS FROM LATE winter to early fall, with the most intense, pronounced, and obvious heats occurring in the late spring. There is much individual variance; some mares have seventeen day cycles while others come in only every thirty days. Maiden mares when first taken to the stallion show signs of heat continuously; it is best to pass them until they establish a regular cycle. While some mares come in heat regularly throughout the year others come in only in the spring when under the stimulating effect of warm weather, increased sunlight, and new grass. The average heat period lasts five days, but may vary from a few hours to as long as a week. There are also split periods with a mare in one day, out the next, and in on the third day again. A record of each mare's estrus cycle should be kept. In some mares the occurrence of heat is obvious even to the untrained observer; "What is the matter with that mare?" visitors to the farm ask. The mare's behavior, which is also called horsing or show, consists of urinating frequently in small quantities; she squats, elevates the tail, opening and closing the vulva spasmodically (winking). Often she raises her neck, stretches out her head and curls back the upper lip. Other mares have so called "silent heat," ovulation occurring with no outwardly visible manifestations. That they are indeed in season can only be determined by examining the cervix with a vaginal speculum or by ovarian palpation, both procedures being performed by a veterinarian, preferably an equine specialist.

As some mares need the stimulus of a male horse to "show," large farms with a great many mares keep a teaser, a well-mannered stallion of no particular value genetically, to induce signs of heat in the mare. It is

Gentle Shetland pony stallion.

imperative that he be kept under absolute control at all times. Walking horse stallions are popular for this purpose because of their easy-going disposition. Many Shetlands are used as teasers because they are easy to handle, easy to maintain, and because, due to the disparity in size, it is unlikely that they would be able to cover a mare accidentally. We use the ridgeling, Buddy, for this purpose. He is so gentle that he can be ridden bareback with halter and shank to check out a mare. Although hormonally a male, he is sterile, obviating the risk that he might breed a mare by accident.

A solid partition between mare and teaser is preferable to a fence. A special wall or the lower half of a Dutch door will prevent injury to the mare, stallion, and handlers.

From a natural point of view, the optimum breeding time is in the late spring when the heat periods are the most pronounced. Winter coats are being shed, indicating a renewal of ovarian function, the green grass is high in protein, and the lengthening daylight has a stimulating effect upon the gonads of both mare and stallion. A late spring conception insures that the foal will be dropped the following year in mild weather when the new grass is growing and before fly season begins. The conception rate (ratio of mares covered to mares settled) is the highest at this time of year. There is an old saying that mares settle on the last service, which is another way

Teasing through a fence.

Teasing over the barn door. (Photo by John Boardman)

of saying that mares which are repeatedly covered, beginning the 1st of February, have a way of catching in May and producing April foals more or less as nature had intended.

In raising Thoroughbreds and Standardbreds, breeders try to have foals dropped as near as possible to the January 1 birthday which all foals share for race purposes. This is of the greatest importance if the youngster is to race at two, as each month of development counts far more at this early age than at a later date. The management problems involved in winter foaling in a severe climate can be considerable. That the foal must be kept confined a greater part of the time may partially offset the advantage of the early birthday.

Within a week after foaling (seven to ten days) the mare comes in season. This period is known as the foal heat, nine-day period, or post-parturient estrus. Opinion is divided as to whether it is advisable or desirable to breed the mare back at this time. It is a superstition among horsemen, not borne out by fact, that a mare is more likely to settle at this time than at any other. It is true that some mares do not show external signs of heat while nursing a foal, but if careful records are kept she may be tried at thirty days and will usually accept the horse. The only advantage to breeding in the foal heat is that it speeds up the reproductive process, next year's foal being dropped almost a month earlier than this year's, an important consideration in a Thoroughbred mare which has produced a late foal. Many famous race horses, however, have had relatively late foaling dates: Man o' War, Capot, Count Turf, Hill Gail, Dark Star, Native Dancer, Northern Dancer, Determine, Hasty Road, Nashua, Bold Ruler, *Gallant Man, and Tim Tam, to cite a few. The danger of infection is far greater in the foal heat; a high percentage of aborted fetuses and abnormal or diseased foals are conceived on the ninth day postpartum. Mares which have produced their first foal should *never* be rebred in the foal heat as there is inevitably more or less parturient trauma to maidens. Mares which have had difficult foalings, which have sustained any injury, which still have a vaginal discharge, which have retained the placenta over three and a half hours should definitely *not* be rebred at this time.

Fillies come in heat for the first time at about eighteen months. The first periods are probably anovulatory, although there are cases of conception occurring accidentally. An occasional colt is also fertile as a long yearling so the sexes should be separated early in their yearling year as pregnancy can only be considered a grave misfortune at this stage resulting in probable stillbirth of the foal and stunting of the dam. A mare should not be bred until she is three or preferably four or it will be at the expense of her own development. A mare bred at three should have a year's rest, not being rebred until she is five. She may continue to be bred throughout her lifetime as long as she maintains an estrus cycle. When she is too old to breed, she will no longer conceive. There is a somewhat greater risk involved in breeding aged maiden mares. The famous trotting mare Goldsmith Maid campaigned through her twenty-first year, producing her first foal at the advanced age of twenty-two; the Morgan mare Polly Forrest continued to produce well into her twenties. Plucky Liege was twenty-three when Bois Roussel was foaled. At Ashforby, England, in 1964, a mare named Cinders, age thirty-one, dropped a foal which was described as "a fine, healthy colt." This was the third foal born in consecutive years after the mare became lame in her late twenties.

In sending a mare away to be bred, it is best to ship her so that she arrives at her destination several days in advance of her next scheduled heat period. It sometimes happens that if she is shipped only *after* she is already in season the excitement of the van ride will put her off and another three weeks, as well as board and vanning expenses, will be lost.

When it has been definitely determined that the mare is in season, her tail is wrapped with a clean two inch bandage. Disposable cotton gauze may be used, or a regular leg bandage which must be washed thoroughly with soap and rinsed and hung in the sun to dry after each use; an Ace bandage may also be used. Starting at the root of the dock, wrap the bandage around a couple of times, spiral down the tail for eighteen inches, twist to equalize tension, spiral back up and tie or pin. Enough pressure should be used so that the bandage does not come loose during service, but not so tight as to cause irritation. Not only does an untied tail act as an impediment to copulation, but the hairs may cut the genitalia of both horses.

The buttocks and external genitalia of the mare are washed with a mild antiseptic soap such as Phisohex. An erection is induced in the stallion and he is washed and rinsed thoroughly. This important phase of management if often overlooked by the backyard breeder as it is difficult to accomplish without an assistant. The spread of infection is greatly reduced by using sanitary measures.

If the mare has a foal at foot it must be confined safely in a box stall so constructed that it is impossible for the foal to climb out or to get a foot caught when he attempts to climb out. All feed tubs and buckets must be removed; there should be no stationary manger, hay net, or any protruding object on which the foal may be hung up or otherwise injured. Many foals have been seriously, even fatally, hurt when the simple precaution of removing everything but bedding from the stall was not taken.

Under no circumstances should the foal be allowed to run loose. In her

Twenty-year-old Morgan mare, Tippy Dee, prepared for breeding. The tail has been wrapped and the buttocks and external genitalia washed. Note the sleek coat, the quality of mane and tail, clean tendons and good bone, and the clear eye, all indicative of soundness and good health, which are of vital importance in breeding stock. Tippy-Dee continued to produce foals until she was twenty-three. (Photo by John Boardman)

concern for his safety the mare would be anything but receptive to the stallion, attempting to fight him off by biting, kicking, and striking. The foal would be certain to be injured in the scramble.

If the mare is stabled only a few miles from where the stallion stands, the foal may be left behind in his own stall; this is the usual practice in Central Kentucky, where there is a large concentration of breeding stock within a relatively small area. If the mare is very excitable, however, it is better to bring the youngster along.

If the mare has showed to the teaser—many farms actually allow the teaser to mount to insure her receptivity—and she is definitely in season, no restraint is necessary, although a shank may be passed over her nose for added control. Breeding hobbles are used to protect the stallion should the mare kick, not to force a reluctant mare to accept the horse. As some mares will kick, in season or not, and a valuable stallion should not be subjected to undue risk, certain breeders routinely use hobbles regardless of the mare's manners and disposition. There is a very real risk that the mare may be cast and trampled, therefore other conscientious breeders use hobbles only on those mares known to be vicious.

If the mare is high-strung it may be necessary to put a twitch on her to divert her mind from the foal and make her stand quietly. Often all that is needed to prevent kicking is for an attendant to take hold of the mare's upper lip as the stallion mounts and dismounts. Within reason, the less restraint employed to protect horses and grooms, the more likely is conception to occur.

The attendant holding the mare should stand behind a barrier of some sort, a simple snorting pole, a solid partition built for the purpose and also used for teasing, or a barn half-door. A breeding stall, a stock-like structure which confines the mare on three sides, is sometimes used; there is risk of injury to the mare should she be cast. In Central Kentucky breeding is conducted in a special shed adjacent to the stallion barn. The ceiling is high and the tan bark floor provides good footing.

The stallion may be controlled by a special stallion bridle. With well mannered horses the shank chain passed over the nose is usually enough. More severe control is effected by passing the chain through the horse's mouth and fastening the trigger snap to the ring at the cheek on the opposite side.

Pasture and corral breeding are dangerous for both mare and horse. In pasture breeding where the stallion is allowed to run with a band of mares the actual date of conception and hence the future foaling date is uncertain. Stallions which run with the mares become extremely aggressive and may attack anyone who sets foot inside the fence. The one advantage to pasture

Chain through the mouth for more severe control.

breeding is that it results in a higher conception rate than hand breeding. In corral breeding the mare and stallion are turned into a small enclosure when the mare is believed to be in season. This practice has little to recommend it as it combines the worst features of both hand and pasture breeding. Occasionally corral breeding is beneficial in the case of a reluctant young stallion which needs time and courage to cover a mare; the mare's disposition must be the best and she must be in heat beyond the shadow of a doubt.

During his first year at stud the young stallion's book should be limited to a few mares. Unless he has been an outstanding race or show horse it is unlikely that the demand for his services from owners of outside mares will be excessive until his get have had an opportunity to prove themselves.

Although some well developed two-year-olds and even an occasional long yearling are capable of siring foals, most breeders prefer to wait until the colt is older, usually when his racing or showing days are over. He may breed one or two mares at two without harm; this is sometimes done so that a high priced young animal may be offered for sale as proven as in the case of the Morgan Vigilmarch which sold in 1959 at the highest price ever paid for a two year old Morgan colt. Three-year-olds may easily cover a dozen mares although they are sometimes given a book of twenty if they are exceptionally promising such as Sir Gaylord, leading three-year-old of 1962, which broke down on the eve of the Kentucky Derby and was promptly retired to Claiborne Farm. Older stallions may serve thirty or forty mares although their engagements are usually limited in their first year at stud regardless of actual age. Draft stallions are able to handle more mares—as many as one hundred—than are stallions of the light horse breeds; Standardbreds are able to cover more mares than Thoroughbreds without ill effect. In 1956, forty-seven Standardbred foals by Gene Abbe were recorded. At the age of nineteen he sired eighty, having gotten almost one hundred in each of the three preceding years! Among the breeds which do not have a sharply curtailed mating season due to racing, the horse may serve a greater number of mares, spread over a longer period of time.[1] In many breeds, however, where culling is not rigidly employed by competition, there are almost as many entires as mares, so that a full book does not exist except with the exceptional horse. There is a feeling among horsemen that the quality of the offspring tend to decline if the horse is used excessively; it is uncertain whether this is due to lowered vitality of the sperm or to lack of mares of uniformly high quality.

If two mares are in heat at the same time, one may be covered in the morning and the other in the evening without imparing the stallion's vigor. In a week of doubles, however, he will benefit from a day's abstinence. If only one mare is in season she may be covered every day of her heat period; if there are several mares in at once, she may be covered every other day so as not to overtax the stallion. Ovulation probably occurs toward the end of the heat period. Opinion is divided as to how long spermatazoa survive in the mare's genital tract: Neumann and Salzman concluded from a series of postmortum examinations that sperm retain their motility for only 28 to 36 hours while Burkhardt observed that they survived as long as 68 to 138 hours.[2] In either case, breeding every 48 hours should give adequate coverage.

Overbreeding, especially in the immature stallion, may lead to temporary impotence, the horse showing complete lack of interest even though the mare is in heat and is receptive. The best cure is a prolonged rest.

If a stallion of any age begins to show diminished interest, fails to settle a large proportion of his mares, or goes back in condition it is probable that he is being over-used even though he covers no more mares than the vigorous rascal next door. The greatest care must be taken with very young and very old subjects.

The age to which a stallion remains fertile is subject to great individual as well as breed differences. Some horses are all through at eighteen while others still stand to a full book. Many stallions remain in active, though limited, service through their twenties. Bimelech was conceived when Black Toney was twenty-seven.

[1] The United Trotting Association, however, permits limited use of artificial insemination which enables a stallion to serve a larger book than a Thoroughbred stallion. The Jockey Club permits the use of artificial insemination only to reinforce natural service.

[2] Pierre Lieux, D. V. M., "Infertility in the Stallion," *Equine Medicine and Surgery* (Santa Barbara: American Veterinary Publications, 1963) p. 633.

4
Breeding Problems

THERE ARE AS MANY BREEDING PROBLEMS AND IDIOSYNCRASIES AS THERE ARE horses at stud.

Stallions may be infertile from any one of several causes, the most common being abnormality of the semen: lowered sperm count, non-motile and abnormal spermatazoa, or too great viscosity. Lowered sperm count may be due to over-use, masturbation, infection, hormone imbalance or dietary deficiency. A sperm count should be done before the horse stands and periodically thereafter if he is not doing well. A stallion with a low fertility rate will profit by regular vigorous exercise such as being galloped several miles a day, exposure to sunlight and fresh air and the feeding of the best pasture and concentrates. Injections of Vitamin B and gonadatropic hormone have been helpful in some cases and antibiotic therapy is indicated in infections, but these are veterinary provinces and it is short-sighted economy for the layman to attempt to diagnose or treat them.

As mentioned above, some stallions are a failure at stud due to impotence or lack of libido. Others, particularly young horses, become impotent temporarily through over-use.

Many stallions have personal preferences and prejudices, performing adequately with mares of a certain color, but showing lack of interest or positive distaste for mares of a different color. Camouflaging the mare under a blanket or sheet is sometimes effective! Other horses are able to perform only in familiar surroundings.

Many vices and foibles date to the first time the stallion was used to cover a mare. It is imperative that the first mare be docile, calm, and in heat beyond doubt, for the young stallion is sure to be awkward in his first attempts to mount. A mare that squeals and kicks will discourage and frighten a timid young stallion. He knows something is expected of him, but is not sure what, where, or how. No roughness should be condoned. He must be punished for biting, kicking, or otherwise savaging the mare. Often several days are necessary before the horse breeds the mare successfully, after which there should be no further difficulties. Stabling him next to a mare in heat or keeping him in a small paddock where he can see and smell her may be of help. Often lack of interest is a sign of immaturity, as puberty occurs at different ages in different breeds and individuals. Similarly, potency declines with advancing age.

When a stallion fails to settle a good percentage of his mares or shows lack of interest although he has not been used excessively, it is probable that he has been masturbating. He induces in himself an erection, sexual excitement, orgasm, and discharge of semen by rubbing his penis against his belly or between his front legs. Stallions are not always observed in the act and those who work around them may deny that they practice this vice but the presence of dried semen between the forelegs is pathognomonic. The horse should be fitted with a hard rubber "stallion ring" designed to prevent erection. These rings come in several sizes and must be fitted carefully so as not to interfere with normal circulation. The rings must be removed weekly to be cleaned and, needless to say, before the horse serves a mare. It is seldom necessary to resort to more severe methods such as shields and cages. Like most other vices, masturbation is most often encountered in horses that are over-confined.

Some stallions do not get mares in foal due to failure to ejaculate. They mount and enter in a normal manner but dismount without discharging semen. As semen is expelled by the rhythmic contractions of the muscles of the genital tract accompanied by switching of the tail, which is known as "flagging," stallions suspected of this habit should be carefully observed and allowed to cover the mare a second or third time if ejaculation is not thought to have occurred on the first service.

VENEREAL DISEASE AND INFECTION

Dourine (*mal du coït*) is a venereal disease of horses caused by trypanosomes. It is transmitted from diseased to healthy horses during copulation. A healthy stallion may pass it on from a diseased mare to a healthy one without contracting dourine himself—hence the need for scrupulous cleanliness during breeding and the washing of the horse both before and after service.

An infected horse loses condition rapidly, exhibits swelling and inflammation of the genitalia, frequent urination, and has variable edematous swellings on the lower part of the body. Hive-like welts appear and persist for days or weeks. The hair in these areas may fall out or become depigmented. The patchy appearance of the coat is suggestive. In advanced cases there is general paralysis. Treatment is not satisfactory and infected animals should be destroyed.

Through vigorous measures such as destroying mares and castrating stallions, the disease has been virtually eliminated from the United States except for several isolated areas of the Southwest near the Mexican border. There was an outbreak in Virginia, however, as recently as 1953; therefore, anyone who stands a horse should require a veterinary certificate with each mare to be bred.

Coital exanthema is extremely contagious. It produces temporary sterility in both mare and stallion. It is characterized by lesions of the genitalia which progress from vesicles to pustules to small ulcers. In the mare there is a vaginal discharge and the affected parts are swollen and extremely sensitive. The disease is transmitted not only by direct contact during mating but also indirectly by contaminated sponges, pails, bandages, or attendants with dirty hands.

The affected areas should be washed daily with a mild antiseptic solution. Cauterizing the lesions with silver nitrate will hasten their drying. The handler should take extreme care not to spread the infection to healthy horses and should rinse his hands with alcohol after first washing thoroughly with soap and water.

Fortunately, coital exanthema leaves no lasting ill effects.

INFERTILITY

The causes of infertility in the mare are more varied and complex than in the stallion. The artificial lives which many animals lead and the unnatural conditions under which they are bred contributes to the low fertility of the domestic horse. Mares that are sick, ill-fed, emotionally disturbed, or suffering from an hormonal, dietary, or metabolic disorder will have a disturbed estrus cycle. Either they do not come in heat or, if they do, they fail to settle.

Mares with silent heat (no external manifestation of estrus) are a real problem for the backyard breeder. Although these mares, in spite of their peculiarities, are as likely to get in foal as any other bred at the right time, they will probably remain barren unless they are of registered stock of the highest potential, since it is a major expense to have a vet out almost every day to palpate the ovaries and examine the cervix with a vaginal speculum.

Certain mares have silent heat periods only when they have a foal at foot but can be bred on the 30th or 51st day after foaling without the expense of having them "specked."

Mares at the other extreme show almost continuous signs of being in heat. If the cause is cystic ovaries, massage and injections of leutinizing hormone are helpful in some cases. True nymphomania is rare; as the ovaries are atrophied, treatment is usually unsuccessful. These mares are impossible to settle and are vicious with other horses.

Maiden mares usually have irregular cycles when first bred, often remaining in heat for a month or so, probably because of local irritation. After they have established a normal pattern they are the most likely to conceive of any group of mares. Stud farm managers prefer to get race mares in late summer or early fall, as they present a special management problem; being drawn very fine, they need to relax, put on weight, and be built up before the beginning of the breeding season. These mares have been under constant tension, more damaging to the female reproductive system than to the male. To prevent soreness, many have received corticosterone injections, which disturb the estrus cycle, and they are a long time in getting back to normal. Some never do.

If mares coming from the race track are merely turned out they go through a let-down syndrome. Sudden relaxation of tension is physiologically as well as psychologically damaging. They actually lose flesh although doing no work, becoming boney and rough-coated. Many become anemic due to their hemapoietic system's having become dependent upon the constant high output of adrenalin during competition.

Ideally, race mares should be let down gradually, being ridden a diminishing amount each day, two hours of slow work being preferable to fifteen minutes of galloping. They should continue to be groomed daily for some time. In this way they may ease into their new mode of life without a sharp transition. When first turned out they should be put into small enclosures to discourage their tendency to "run their heads off."

Often it is not feasible to work retired mares, as they are too sore or crippled to be ridden. A typical case was Dot Sena, which we acquired in December of her six-year-old year after she had broken down. As she could scarcely walk, any but the mildest exercise, such as spending the daylight hours in a small paddock, was out of the question. She was a poor doer eating only a handful of grain at a time. By spring she was actually thinner than when she had come out of training. She was sent away to be bred, was in heat continually for thirty days and failed to conceive. At the end of a twenty-one day interval she again came in season and was rebred success-

Dot Sena winning a close decision. The tension under which Thoroughbreds race is evident in this picture. (Photo by Paul Anderson)

Retired race mares let down and enjoying a rest in the late fall sunshine.

Dot Sena, an eight-year-old Thoroughbred, and Tippy Dee, a twenty-year-old Morgan, both about eight months in foal. Brood mares benefit by regular exercise. (Photo by John Boardman)

fully. By this time, as her lameness had completely disappeared, she was able to have daily work under saddle. Her appetite improved until she was cleaning up eight quarts a day and looking for more. Later, her grain ration was sharply reduced when it seemed that she was in danger of becoming too fat!

Overly fat mares are notoriously poor breeders. Reduced rations and regular exercise are usually the only measures needed to get them in the proper breeding condition.

Many horsemen have great faith in wheat germ oil, which, when fresh, contains Vitamin E. It is given at the rate of one ounce twice a week to maiden or problem mares, beginning in January and continuing until the mare is believed to be in foal. Poured over the grain it is greatly relished by most horses.

Infections of the genital tract (see coital exanthema) may render the mare permanently or temporarily barren. Mares bred in the foal heat are not only less likely to conceive than at other times, they are more likely to become infected. Lack of sanitation at the time of service or disease of the stallion are other causes of infection in the mare. The use of an unsterilized vaginal speculum may transmit germs from one mare to another. The dubious practice of "opening" a mare before breeding has made many mares barren by introducing infective material into the cervix and uterus. Artificial insemination may cause irritation and infection if hygienic measures are not scrupulously observed. Some mares have a peculiarity of conformation which predisposes them to genital tract infection. A thin-walled, oblique (rather than vertical) vulva is predisposed to windsucking, whereby feces are drawn into the vagina causing infection which may extend to the uterus. To prevent contamination and infection, windsucking mares undergo Caslick's operation; the lips of the vulva are sutured together; the sutures must be removed prior to breeding and again prior to foaling.

Mares with an infection of the genital tract usually have a vaginal discharge. They may or may not have a slight elevation in temperature. They should be examined by a veterinarian, a culture made of the discharge, and the proper treatment instituted. Usually a broad spectrum antibiotic is given both locally and by injection. The mare should not be bred until the infection has completely disappeared.

Another cause of sterility is mechanical impediment to the deposit of semen in the uterus. Injuries or long standing infection may have sealed the cervix with scar tissue. Mares which have been sutured for windsucking (see above) cannot receive the horse until the stitches have been removed. Persistent hymen is rare but is an occasional cause of barrenness in maidens;

the hymen may be exercised by a veterinarian, never opened by a groom with unwashed hands! The uterus may be filled with cysts or tumors which make pregnancy impossible. Occasionally the fallopian tubes are occluded, preventing the normal descent of the ovum into the uterus; this can be diagnosed by trying to blow air through the tubes, a veterinary procedure.

Some mares are barren due to psychological causes; mares which are unhappy or keyed up fail to conceive. Old age and extreme debility render a mare barren. Either she fails to come in heat or, if she does, she fails to conceive—no doubt a protection against demands on her reserves of strength which she would be unable to meet successfully.

Idiopathic sterility. A small percentage of mares remain barren when they appear to be normal in every other respect, just as certain stallions are sterile although their sperm count would seem to indicate a high degree of fertility.

ABORTION

If the foal is dropped before it is viable it is said to have been aborted. There are many causes. An ovum of low vitality fails to develop properly, dies, and is passed out. If the uterus is covered with tumors or cysts, implantation is impossible and the embryo is expelled. An hormonal imbalance will result in abortion, especially at the critical period around five months when the manufacture of progesterone is taken over by the placenta instead of the ovaries. Mares which habitually abort at five months can be treated successfully by injections of progesterone. As a high percentage—an estimated one third—of aborted fetuses are abnormal in some important respect, abortion functions as a form of quality control by eliminating at an early stage most of those individuals which are incapable of extra-uterine existence. Twin foals are frequently aborted or delivered prematurely. Viral infections of the mare invade the fetus. *Brucella abortus*, the causal agent in Bang's Disease (contagious abortion) in cattle, and Brucellosis (undulant fever, Malta fever) in man also causes abortion in horses. Although the disease is no longer common in this country, it should be suspected in otherwise unexplained abortions in mares, particularly those with fistulous withers and poll evil, or which have been stabled with cattle, pigs (*Br. suis*), or goats (*Br. melintensis*).

An estimated 5 to 10 per cent of all abortions are caused by mycotic placentitis. The portion of the placenta adjacent to the cervix is first affected; the fungus, which grows slowly, invades and destroys the villi through which the foal receives oxygen and nutrients. When sufficient

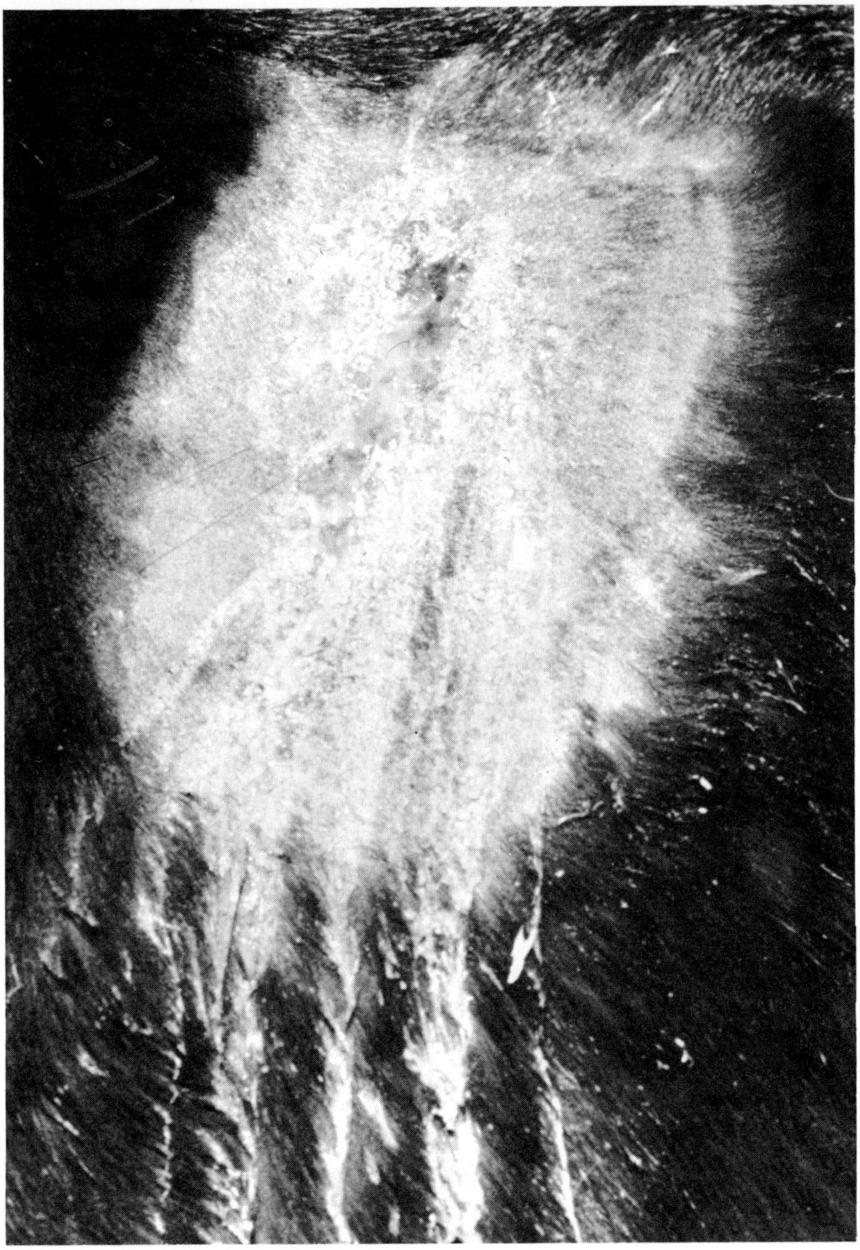

Fistula of the withers. (Courtesy of Ohio State University)

damage is done, the fetus can no longer be maintained; it dies and is slipped. Occasionally the destruction of placental tissue is not great enough to cause abortion, but stunts the foal, which is born small, weak, and underdeveloped. Mycotic placentitis is thought to be due to failure to observe the usual sanitary precautions at the time of breeding.

Occasionally the umbilical cord becomes strangulated, cutting off the fetal blood supply. Falls, kicks, blows, hard use, and accidental breeding may terminate pregnancy prematurely. As a rule, however, the mare can undergo a great deal of abuse without slipping her foal, so well is it protected by its membranes and cushions of fluid. In the eighth month of pregnancy Dot Sena was clowning, hit a patch of ice, turned a sommersault, landed on top of a fence plank, and tumbled into an adjoining paddock. Not only was she severely bruised on the lower abdomen but the pectoral muscles were lacerated through to the sternum. (See photos in Chapter 12, "Injuries.") Intermittent contractions, vaginal discharge, and bagging up were observed for a week or more, but the mare did not abort. About ten days before her earliest possible due date, she produced a colt perfect in every respect except that he was pony size. Beverly Blue received a kick on the abdomen severe enough to raise a grapefruit size hematoma without ill effect to the filly she was carrying. An Arabian mare was reported to have been struck by a car in the ninth month of pregnancy and, though she sustained severe lacerations and several broken bones, she carried the foal to term, both the delivery and the foal being normal.

In early abortion there is no clinical sign other than that the mare comes in heat after an anestral period during which she had been presumed to be in foal. At four and a half months Replevin slipped her foal following an outbreak of viral rhinopneumonitis. At this stage there was relaxation of the external genitalia, and a thick, reddish brown discharge lasting a week, after which the mare came in season. The udder enlarged but not to as great an extent as at term. The mare produced milk, took a very motherly interest in all the young foals and was affectionate and craving of attention from all of us. Still farther along in pregnancy the signs of abortion are the same as actual labor—the mare lies down, there are violent contractions, the foal is expelled but is too immature, unformed, and weak to survive.

5
Diagnosis of Pregnancy

AN ACCURATE DIAGNOSIS OF PREGNANCY IS OF THE UTMOST IMPORTANCE TO A successful breeding operation. With high stud fees, board bills, transportation costs, and long gestation period, it is essential that the owner of the mare be able to determine whether she is in foal or barren.

Cessation of the estrus cycle is a presumptive but not an infallible sign of pregnancy. The mare regular in her heat periods which refuses the horse or does not show to the teaser at twenty-one days following the last service may be assumed to be in foal, although this is not always the case. Some mares horse only in the warm spring months. On the other hand, some mares show signs of heat while pregnant, are bred, and often slip their foals as a result.

Between forty-five and one hundred days following the last cover a Friedman test may be performed. Blood withdrawn from the mare is injected into a virgin rabbit. In a positive test result, the rabbit's ovaries show enlargement due to the presence of gonadatropic hormone circulating in the mare's blood stream. The Friedman test is considered to be 99% accurate.

After 120 days a Cuboni or urine test is valid; it is 90 per cent accurate.

A competent veterinarian can make an accurate diagnosis of pregnancy by rectally palpating the uterus and ovaries, noting the consistency, size, and contour of the organs. There are also characteristic changes in the cervix as seen by a vaginal speculum.

To someone who knows the mare well and is familiar with her disposition, a change in temperament is indicative of pregnancy. A high-strung mare may become calm or a placid mare greedy and aggressive. Many mares are very affectionate, especially toward the end of pregnancy. Other mares become sluggish and dull.

After seven months the measurement taken at the waist is greater than the heart girth. Even in full-bodied animals the waist circumference is normally a few inches less. This method is not accurate especially in multiparous mares. Some mares show few outward signs that they are in foal at all until the last month of gestation. Beverly Blue assumes the shape of an obviously pregnant mare only in the tenth month, thereafter growing rapidly and becoming visibly more ponderous each day. There have even been embarrassing cases of mares performing faultlessly over fences in a hunter class, returning to the stable area, lying down and giving birth to a hitherto unsuspected foal!

From six months, the movements of the foal can occasionally be seen in the area of the mare's flanks or near her naval. The fetus is liveliest when stimulated, as when the mare takes a cold drink of water or when she is slapped, gently, on the side.

After the foal has reached a sufficient size, ballotement is another diagnostic method. On pushing on the mare's flanks the examiner feels a mass—the foal—which floats away and then rebounds to the palpating hand.

6
Foaling

THE GESTATION PERIOD IN THE MODERN HORSE IS SLIGHTLY OVER ELEVEN months. Live foals have been dropped at 300 days and some have survived which were carried until the 400th day.[1] Some mares have a longer gestation period than others. The Thoroughbred mare Ethel M. habitually carries her foals a full year. Winter foals are carried longer than summer foals and colts longer than fillies. A Russian report based on data obtained over a 100 year period in Thoroughbred and trotting studs indicates that mares with a gestation period longer than the breed average not only lived longer than the average but produced high quality offspring when over twenty, continuing to breeed to the age of twenty-five or twenty-six. The gestation period lengthened with each pregnancy. Mares with a relatively short gestation period, however, were less fertile and did not produce outstanding foals beyond the early years nor were they themselves long-lived. In this group there was a tendency for the gestation period to become shorter with each pregnancy.

It is important to keep a careful record of each date the mare was covered as well as the duration of her previous pregnancies and any peculiarities which she may have.

About a month before foaling bagging up begins. Day by day the udder becomes larger, rounder, fuller, and more firm. If the mare is stabled at night, the bagging up is most pronounced in the morning before she is turned out, and may have disappeared entirely by evening. Mares kept out of doors all the time show less sign of bagging up than stabled mares. Change in the udder is more pronounced and is noticeable earlier in maiden than in multiparous mares, as the udder never quite regains its former shape

[1]The pony mare, Annie K., was bred to the Morgan stallion, Blackacre Mandamus, on May 12 through May 15, 1969 and produced a live foal on July 26, 1970, a gestation period of 437 days. This may be a record.

Morgan brood mare, heavy in foal.

Mare immediately prior to foaling. Note depression at root of tail and swellings extending forward from the udder on the lower portion of the abdomen.

after pregnancy; transient swelling may be noticed as early as three months before the foal is dropped, but a full udder earlier in pregnancy is an ominous sign indicating that the mare is about to abort. A week before foaling a yellowish, waxy secretion appears on the teats; this is called "waxing up." The wax increases, gumming the inner side of the hind legs. About forty-eight hours prior to foaling, milk begins to drip, increasing to a stream in mares which are heavy milkers.

In some mares a swelling extends from the udder foreward along the underside of the belly. It may reach as far foreward as the mare's brisket, the circumference being that of a man's arm or even larger. This swelling is caused by pressure of the gravid uterus as it sinks lower in the abdomen just prior to foaling. The swelling is absorbed, disappearing a few days after foaling.

Throughout pregnancy the uterus contracts from time to time. The contractions become stronger, more frequent, and more obvious to the casual observer in the last month of pregnancy and are often mistaken for actual labor. As gestation nears its end these contractions become painful; the mare may stop eating, stamp a hind foot and look around at her side. It is the passing, isolated nature of these preliminary contractions which differentiate them from the rhythmical, progressively more frequent and more intense pains of actual labor. Gradual relaxation of the cervix takes place in preparation for foaling.

Each day of pregnancy the abdomen becomes more full and rounded especially in its lower rear portion. Shortly before foaling, as the fetus approaches the birth position, the mare falls in at the flanks and at the root of the tail as the pelvic ligaments soften. The external genitalia relax; the vulva often becomes level with the point of the hip. There is a mucous discharge, quite heavy in some mares, which soils the tail and hind legs. The mare may walk with difficulty and with a straddling gait. She passes manure and urine frequently and in small quantities due to pressure of the gravid uterus on bladder and rectum.

Just before foaling many mares refuse to eat. Others, however, continue to munch between contractions throughout labor up until actual foaling. As the tempo of the contractions increases the mare becomes restless. She walks, paces, kicks, paws. She may bite at her sides, sweat, swish her tail, lie down, get up, roll, rub violently against the walls of the stall and give every indication of having a severe case of colic. The bag of waters appears; with the next pain it breaks. If the mare is not already down she lies down at this point. In normal deliveries a few violent contractions are followed by the foal's forefeet, one slightly in advance of the other, still encased in the membrane. If the hooves point toward the mare's hocks and the foal's

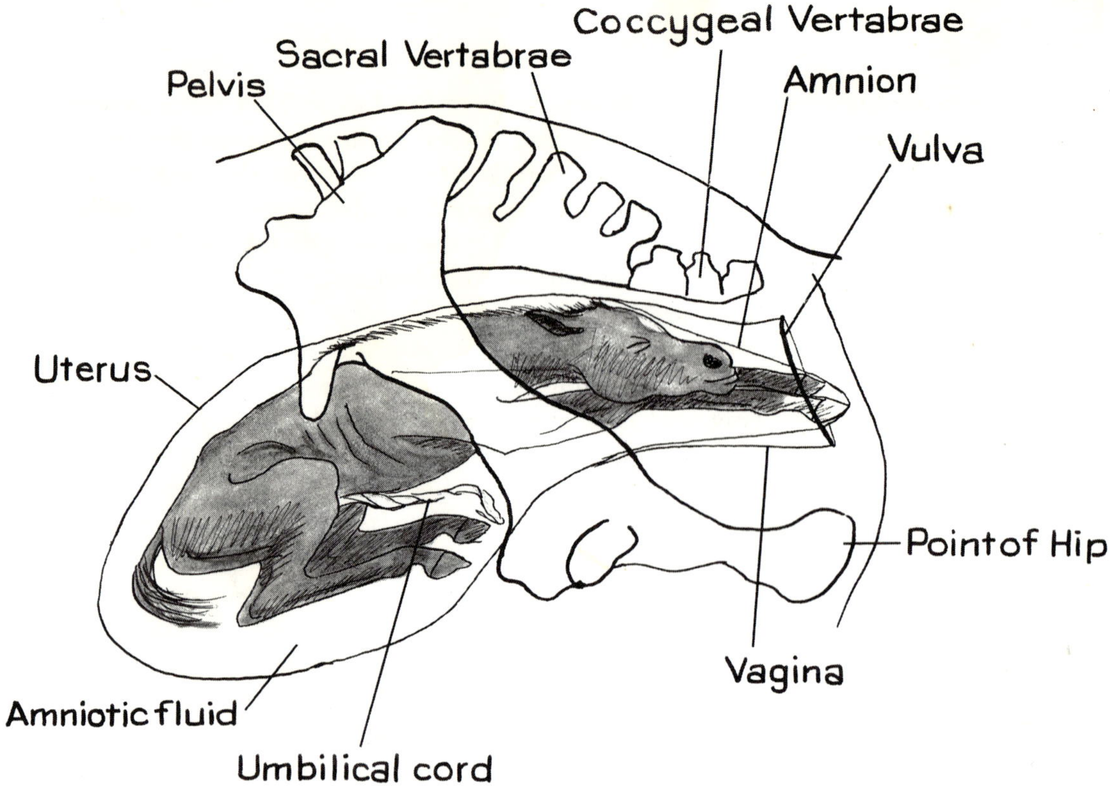

Foal in normal birth position.

nose lies on the forelegs above the fetlocks all is as it should be and the foal will be delivered without help in a couple of minutes. If the foal is very large it may be hung up temporarily at shoulder or hips. Grasping the fetlocks, pull alternately on right and left toward the mare's hocks. Never pull on both legs at once as the idea is to slide one shoulder ahead of the other so that the foal may slip through the pelvis more easily.

If the mare strains for fifteen minutes and no feet appear, if only one foot comes out, if the hooves are pointing toward the mare's tail—indicating that they are hind feet rather than forefeet—something is wrong and a veterinarian must be called AT ONCE. Manipulation of limbs should not be undertaken by an untrained person, as more harm than good is sure to be done, with fatal outcome for the foal and quite possibly for the mare, too. Attaching a block and tackle to the feet is inhumane in the extreme; it will surely kill the foal and rupture the mare if the foal's head is turned to one side or if a leg is bent under.

It is not within the scope of this work to discuss the various malpresentations and their methods of version. It is possible in some cases to manipulate and rearrange the foal's limbs so that it may be delivered alive. Often, because of the sheer mechanical difficulty of working within the inaccessible area of the mare's vagina and uterus and in order to save the mare, which may be near exhaustion, it is necessary to remove the foal piecemeal. In most cases the foal is already dead or dying. Embryotomy is preferable to Caesarean section, which is impractical under field conditions, the mare usually dying of shock.

Ninety per cent of all foalings are uneventful. An additional 5 per cent can be manipulated successfully. Only 5 per cent end fatally for either mare or foal.

Most mares are shy about foaling and will accomplish the act during the five minutes that the attendant decides that nothing is doing and goes into the house for a cup of coffee. Most foalings take place during the early morning hours.

If an attendant is present at the actual birth—and he should make every effort to be on hand—he should break the membrane and dry the foal's nostrils with a sterile towel. Dry the body briskly to prevent chilling and to stimulate the foal. With foals that do not breathe spontaneously, work the front legs briskly back and forth to expand the lungs. The attendant may also breathe into its nostrils.

Never under any circumstances cut the umbilical cord. The foal is still receiving placental blood through the cord and if it is cut before it stops pulsating the foal is deprived of much of his blood. He will be greatly weakened and less able to resist infection. The cord will break by itself when the mare stands.

When the cord breaks, treat the stump with iodine IMMEDIATELY to prevent infection and to hasten its drying. The technique for this is to fill a small container such as an ounce measure or a shot glass with 10 per cent iodine and immerse the stump in the liquid. The cord may be more thoroughly soaked if the foal is rolled over on his back. Never tie the cord as this practice promotes infection and delays the drying of the stump.

Except for drying the foal and treating the stump of the cord do not disturb the mare and foal. Do not encourage the foal to rise. Allow both to rest and recover from their ordeal. When the foal first tries to stand he will be very awkward, flopping around and falling many times. This process helps straighten and strengthen his legs and tone his muscles; it should not be interferred with. The stall is deeply bedded to prevent injury. Some vigorous foals are on their feet and nursing by the time they are fifteen minutes old. Others, usually the longer-legged, less compact types, may be as much as three hours finding the right combination. If the foal shows no inclination to rise after a reasonable length of time there is probably something wrong with him.

This foal is a few minutes old. The mare is still resting.

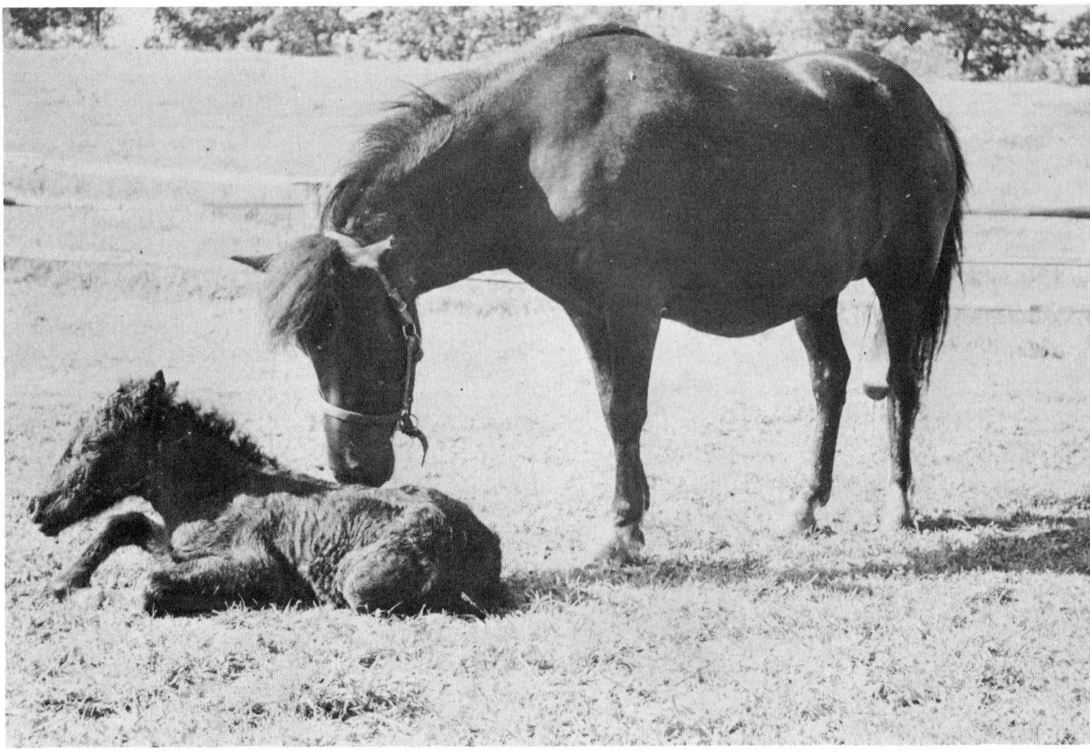

Mare encourages foal to stand.

Foal tries . . .

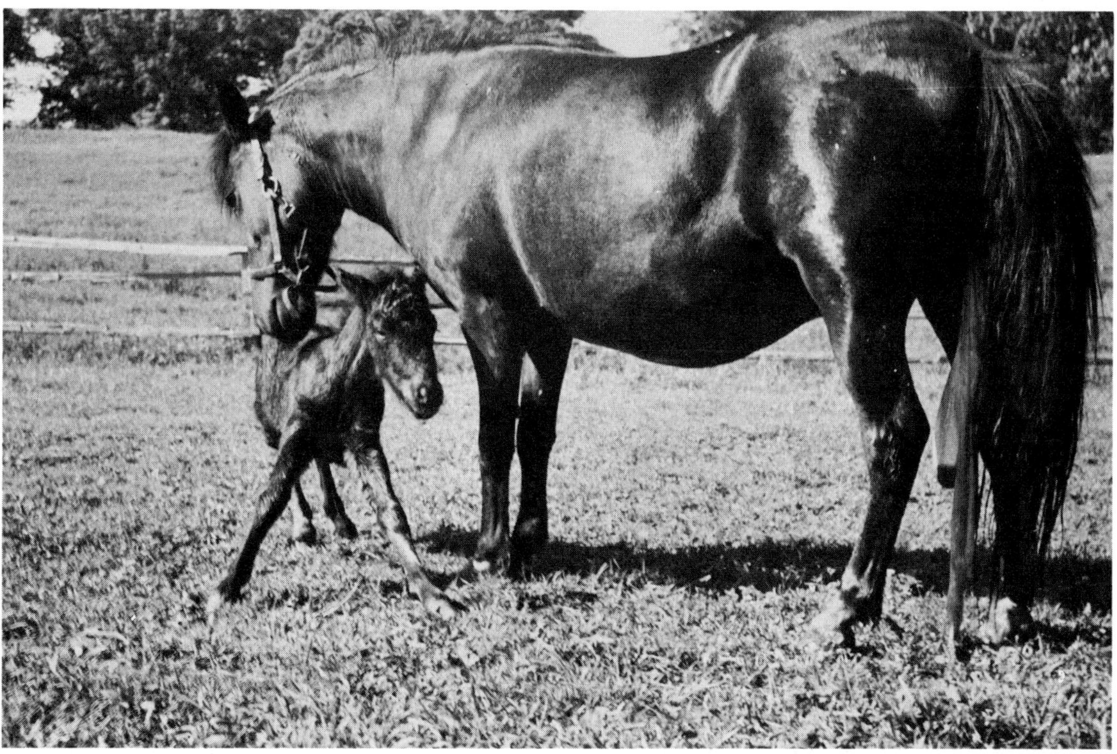

And just makes it.

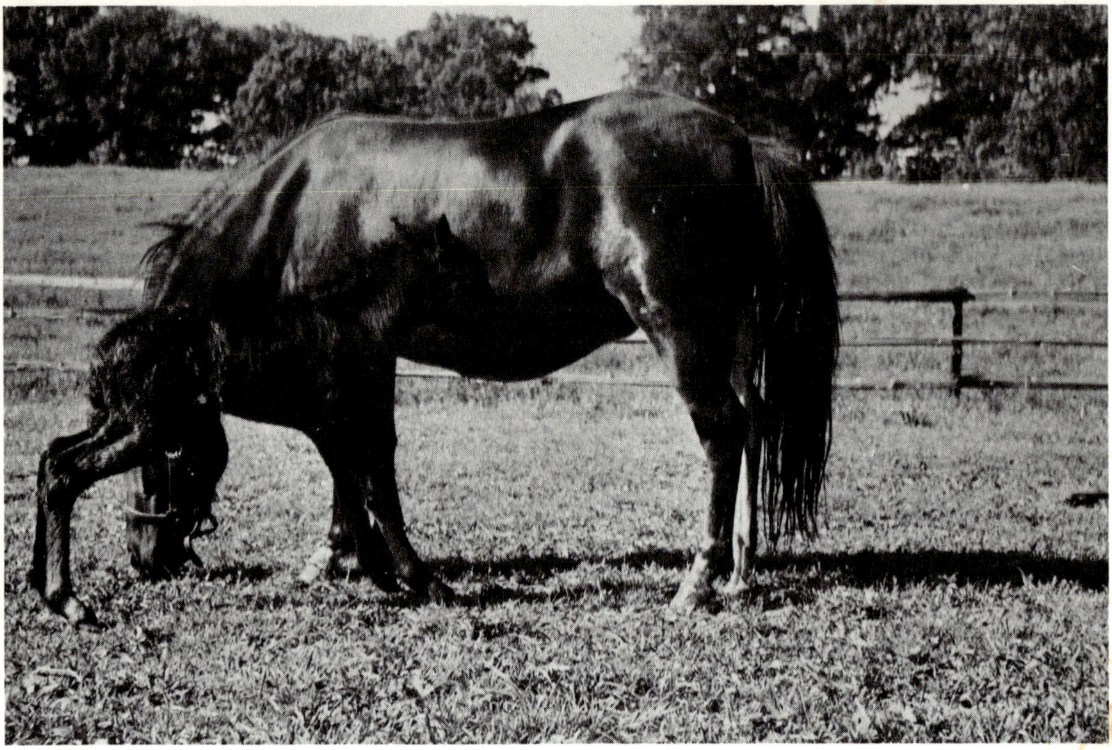

Legs are still very wobbly, but she's headed in the right direction for nourishment.

Within fifteen minutes to two hours after foaling the placenta is usually passed. The mare may lie down or she may deliver it standing. Some mares will not allow the foal to nurse until the placenta has beeen passed. It should be examined carefully to make sure that none of it has been broken off and been retained in the uterus. There will be a tear where the foal came out, but any other tears should be regarded with suspicion. In questionable cases the placenta should be saved for the veterinarian to examine. If the placenta has not been passed within three hours a veterinarian must be called, as the consequences of a retained placenta are serious. The mare may die or be permanently barren if she does not receive prompt attention.

Immediately after foaling the mare may be given a bran mash made by adding boiling water to bran to the consistency of oatmeal and allowing the mixture to steam until cool enough to eat. The addition of linseed and chopped carrots will heighten her enjoyment of the mash. Many successful farms routinely give one gallon of mineral oil by stomach tube to prevent impaction colic due to bruising of the colon during labor. Only a veterinarian should attempt to pass a stomach tube, but oil may be safely added to the mash. The mare is fed a laxative diet for several days and her stools are checked closely for at least a week, both for consistency and quantity. Scant, dry manure is a danger sign that a state of impaction is developing.

If the mare has sweated during labor she should be rubbed dry and blanketed in severe weather.

The membranes, placenta, and all wet, soiled bedding should be removed from the stall at once and fresh bedding put down.

Maiden mares are sometimes reluctant to accept their foals. They should not be left alone together until the foal is nursing. Many foals have been kicked and trampled to death by their dams because there was no one around to give the mare help and encouragement.

If the mare shows any indication of kicking, hold up a front foot until she will allow the foal to nurse without protest. In extreme cases it is necessary to apply a twitch and back the mare into a corner, but such methods should only be used as a last resort. As a precaution during pregnancy the udder should be washed and massaged to accustom the mare to having it handled. If there is any soreness and congestion present it may be relieved by applications of towels soaked in a very hot epsom salt solution and wrung out, followed by massage with vaseline or mineral oil.

It is of the greatest importance that the foal receive the colostrum, or first milk, which is laxative and helps clear out the meconium, that black, tarry, sticky substance which fills the foal's intestines at birth.

The mare and foal are kept indoors for two or three days if the weather is anything short of ideal. Then, if it is not raining or windy they may be turned out for a few hours in a small paddock by themselves. It is best to avoid the chilly early morning and evening hours in winter as well as the blazing hot noon and afternoon in summer. Most mares are fiercely protective of their offspring during the first week to ten days and will attack another horse if they are turned out together. Mares which have been kindly treated are willing for the owner or other familiar person to touch the foal but will turn their backs on a stranger, shoving the foal into a protected corner. While Tippy Dee welcomes members of our family in the foaling stall, allowing us to handle her new baby, she will tolerate strangers only if they are properly introduced; she watches anxiously while strange children pet the foal as long as one of us is around. When a neighbor came to show the filly to *his* friends, however, the old mare put herself between the baby and the intruders, laying back her ears, cocking a hind leg, and switching her tail in such a fierce manner that they were frightened away, as was her intent.

After a few weeks, if the mare and foal are both doing well, they may be turned out with the rest of the brood mare band. A few mares will continue to be excessively protective, especially hot-blooded mares and mares with their first foals. When Night Season was at foot Beverly Blue would not let him stray a step from her side. He was not allowed to play with the other foals and they or their dams were chased if they made the mistake of coming too close. With her second foal Beverly was more relaxed and, while keeping a close watch, allowed her the freedom to run and play with the other youngsters. Tippy Dee, on the other hand, having raised eight foals and being of a calmer nature, is a very permissive mother after the initial protectiveness has worn off. She keeps an ear cocked for real trouble, however, as when a pack of elkhounds had Replevin cornered in a remote part of the pasture with the obvious intent of making a meal of the filly. The old mare charged to the rescue and sent the dogs flying with tooth and heel. She ignores her offspring's heartbroken whinnying, however, while she goes off on a trail ride or gives a lesson in the ring.

PREPARATION FOR FOALING

Care of Foaling Stall:

All bedding should be removed between occupants, and the floor and walls to a height of four feet thoroughly scrubbed and disinfected.

Bed deeply with straw.

Close out all drafts in severe weather.

Remove manure immediately, as soon as it is dropped.

Preparing the Mare:

Wash dock, buttocks, and vulva, being careful not to touch the vaginal orifice. Use mild antiseptic soap such as Phisohex. Wash hind legs.

Bandage the tail.

Wash as needed any parts which become soiled.

Attendant must wash hands and rinse with alcohol before examining mare.

Supplies to Have on Hand in Advance:

Sterile towels. Towels may be sterilized by sealing them in a paper bag and placing in an oven at 225 degrees for half an hour.

Sterile drop cloth. (Usually not needed, as the foal has a way of being dropped on soiled bedding while the attendant is up at the house having that cup of coffee!)

10% iodine. (Also called "strong iodine" as opposed to the 3% tincture for human use.)

2-inch wide tail bandage.

Antiseptic soap.

Warm water.

Alcohol.

Heat lamp to warm stall in very cold weather. Infra-red chick-brooder bulb does very well.

7
The Newborn Foal

WITHIN TWELVE HOURS AFTER IT IS DROPPED THE FOAL SHOULD BE EXAMINED by a veterinarian. He will give the foal a shot of antibiotics and tetanus, check for any evidence of defect or disease, examine the placenta to make certain that no part of it has been retained, and check the mare for any injury which may require repair. He may give the foal an enema if he thinks it is indicated to assist in removing meconium to prevent impaction.

Good care of the mare during pregnancy and clean surroundings at foaling are the best insurance of a healthy foal. There are many diseases of the newborn which are thought to enter the system through the untreated naval. They are due to filthy surroundings and unhygienic techniques. Weak foals are more subject to infection than vigorous ones.

Diarrhea, with watery stools and signs of abdominal pain, is extremely serious, as the foal becomes rapidly dehydrated. Scouring may be caused by a microorganism (such as *actino-bacillus equi, salmonella, streptococcus,* etc.). There are also diarrheas of viral origin. A vet should be called at the first sign of diarrhea, as young foals often pursue a precipitous downhill course. As a rule there is not sufficient time to prepare a culture and identify the causal organism, the foal either dying or recovering by the time the result of the culture is known. Therefore a broad-spectrum antibiotic preparation is administered every twenty-four hours. Electrolyte solutions are given intravenously in cases of severe dehydration.

Some mares are extremely heavy milkers and the foal scours due to an embarrassment of riches. Care must be taken in feeding to reduce milk production, keeping a fine line between scouring the foal and letting the mare go back in condition. In extreme cases the mare should be partly milked out by hand several times a day. We have yet to raise a foal from

Tippy Dee without its scouring due to her heavy milk production. Her grain is kept at a minimum during the first weeks until the foal can accommodate the generous supply.

Foals scour when the mare is in heat.

Many young foals eat manure, which causes diarrhea. This disgusting habit is fortunately soon outgrown. It is thought to be associated with a mineral deficiency. The only immediate remedy is to remove manure from stall and paddock as soon as it is dropped.

Constipation is as dangerous to the newborn as diarrhea. If the meconium is not evacuated it hardens and impacts as much as two or three feet of intestine. The enema administered shortly after birth guards against constipation as does making certain that the foal receives the colostrum. After the foal has nursed, meconium is replaced by yellow stools. With the addition of hay or grass and grain to the diet the manure becomes darker.

The constipated foal strains to evacuate. He may show signs of colic, stamping or kicking with a hind foot and waving his tail. The balls of manure are hard and dry. Mineral oil or soapy water enemas are used to soften and help remove the semi-impacted mass. Great gentleness must be used as the infant rectum is delicate and may be ruptured by the enema tube. A vet must be called if symptoms persist or if the enema fails to produce results.

Joint-ill or *naval-ill* is caused by a generalized invasion of the foal by bacteria (*Actinobacillus equi, Shigella equirulis, Bacillus nephritites equi, Shigella viscosa, Sh. equili*). The foal is weak at birth or becomes so within forty-eight hours. It either lies in a comatose state—the so-called sleeper syndrome—making only feeble movements, or wanders aimlessly, unable to nurse—the dummy syndrome—collapsing into a coma from weakness. High temperatures are typical; the foal deteriorates rapidly, dying within twenty-four hours. Older foals show arthritic symptoms with lameness and painful joints.

It was formerly believed that joint-ill was caused by a prenatal infection. Prenatal infections are actually quite rare. *A. equili* is acquired after birth with weak foals being the most likely to succumb; that foals which are victims of joint-ill are sub-standard from birth is cause rather than effect of the disease process.

Unremitting efforts toward cleanliness in the foaling stall, sterile technique on the part of the foaling attendant, prompt treatment of the naval, the routine administration of antibiotics, and careful observation during the first hours and days of life are of the utmost importance in preventing and controlling infection in the newborn.

The mare that suffers from cervitis or metritis probably will abort; she

Newborn foal lifting hind leg, indicating abdominal pain.

may produce a diseased foal whose symptoms are similar to that of the joint-ill syndrome.

Pneumonia is most likely to develop in debilitated foals, premature foals, or those dropped in inclement weather or adverse circumstances. Foals with low vitality, which do not nurse readily, or which are weakened from exposure, prolonged labor, or from other disease are the likely candidates for pneumonia.

The symptoms are those of depression. The foal stops nursing; he lies down most of the time or stands listlessly with head hanging; he has an elevated temperature of 104–106 degrees, he coughs, breathes with difficulty. Prompt treatment is necessary if the foal is to be saved. Antibiotics are started at once, often intravenously to hasten their effect. The foal must have good nursing and be protected from drafts and sudden changes in temperature until he is fully recovered. If he is too weak to nurse he must be fed intravenously, the mare being milked out by hand to prevent her drying up. Convalescence will last several weeks during which the foal should be encouraged to eat a nourishing diet.

In general, then, the danger signs are diarrhea, constipation, listlessness, general prostration, fever, and, especially, loss of appetite. Often the only sign that something is wrong with the foal is a full udder in the mare indicating that he has stopped nursing. It cannot be stressed too strongly that no time must be lost in getting professional help. Every hour lost lessons the chances of saving the foal. The newborn has meager reserves of strength; illness often pursues a fulminating course with the foal fading visibly in a brief time. Strong supportive measures followed by diligent and meticulous nursing are needed if the foal is to recover.[1]

The first and best food for the foal is its dam's milk. Should the mare die in labor, milk out the udder and feed the colostrum to the foal from a baby bottle. Orphan foals may be fed a formula of four ounces of evaporated milk to four ounces of water and one tablespoon of Karo. Feeding an orphan foal by hand is a discouraging undertaking. As it must be fed every hour for the first week and every two hours for the second week, the problem of keeping the foal alive can occupy the energies of an entire household. Large farms maintain mares of inferior quality as wet nurses should there be registered orphans to raise, the nurse mare's own foal being destroyed or weaned early. Some draft mares are able to support two foals. Onestep, stakes-placed winner of thirty-three races,[2] was raised by a work horse mare bought at auction for that purpose when his dam, Our Bessie, died shortly after foaling. Thoroughbred foals raised by cold blooded mares are often rank and ill mannered, as draft mares are not as strict disciplinarians as Thoroughbred mares!

When the foal is a few days old he will begin chewing blades of grass or pieces of hay. Experimental, imitative nibbling and tasting precedes actual eating. Some foals begin to take grain from the mare's feed box at three or four days; most are eating grain readily by a month, while others show little interest until later, especially the foals of heavy milkers. When the foal is a month or two old he may be given his own feed tub and the mare tied up so that she will not eat his grain as well as her own. By weaning time he will be cleaning up four to eight quarts of oats a day. Many large farms feed sucklings from "creeps," which are long, common grain troughs surrounded by a low bar under which the foals can easily walk but which keeps the mares out. The youngsters are thus encouraged to eat as much and as often as they wish for rapid growth and development.

If the foal is to amount to anything at maturity, grain must be of the best quality. When the pasture is not plentiful it must be supplemented with hay. In areas where the soil is known to be deficient in certain minerals steps must be taken to correct the imbalance before the land is seeded for a crop. If most of the feed is purchased and if the management program on the farm where it is harvested is unknown, it is wise insurance against the development of vitamin and mineral deficiencies to feed a supplement to mares and foals.

Within reason, horses are meant to be out of doors; winter foals are dropped with dense coats to protect them from inclement weather while summer foals have shorter, sleeker hair. Some foals lose their baby hair

[1] There are several commercial milk substitutes available through your veterinarian.

[2] Onestep won six races at the age of eleven!

during the first summer while others retain it until late in their yearling spring. Contrary to popular opinion, the date of foaling in no way affects the quality of the adult coat. It is an old wives' tale that late summer and fall foals are rough-coated throughout their lives. A rough coat may be attributed to poor breeding (cold blood) parasite infestation, neglect, ill health, or general debility, but not to the month in which the individual was dropped. Horses foaled in the northern hemisphere when exported to India, South America, Australia or South Africa soon reverse winter and summer coat in accordance with local climate.

Due to prolonged folding within the cramped space of the uterus many foals are born with crooked legs. Hind legs tend to correct themselves in a short time. Forelegs straighten, too, but not as rapidly or as completely. When Blackacre Sam was foaled his off hind fetlock was cocked, or overflexed. Within a week it was normal. Debt, at the opposite extreme, had over extended fetlocks and was literally walking on her pasterns. There was also excessive angle at the tarsus (sickle hocks). In a couple of weeks she was ideally straight. Soleil d'Or, Sam's brother, toed out and was over at the knee in one foreleg; at five months it was debatable which leg used to deviate from the straight and perpendicular. Legs which are extremely crooked and which do not correct themselves spontaneously can be placed in casts. A new technique developed by Drs. C. D. Heintz and D. D. Delahanty uses stainless steel staples driven into the epiphysis of the affected limb to correct angular deformities. As most deviations correct themselves within the first couple of months this rather radical treatment (which is performed under a general anaesthetic, preferably in a fully equipped veterinary clinic) is used only in cases which appear to becoming increasingly severe with permanent deformity as a probable result or in those cases which do not respond to more conservative treatment. Diagnosis is confirmed by x-ray; the cases which are selected for stapling are those which show an abnormal epiphyseal plate with overdevelopment on one side and compression on the opposite aspect. Progress is checked frequently by x-ray throughout the 19 week period that the pins are in place. The staples are removed under general anaesthesia. In cases treated between 3 and 9 months when there is still a good potential for growth the prognosis is favorable. Due to the great expense involved stapling should be attempted only with racing stock of the very greatest promise.

8
Early Training

THE FOAL'S TRAINING BEGINS AT BIRTH. WHEN HE IS KINDLY BUT FIRMLY handled from the very beginning he will be easy to break when the time comes. The newborn has no fear of man. In fact there seems to be some doubt in the foal's mind as to which is mother, the mare or the attendant. Most foals, however, will have become quite shy by the following day and will hide behind the mare when anyone comes into the stall. This natural caution is often encouraged by the over-protective mare that shoves her offspring into a corner shielding it with her own body. Patience, plus the foal's own insatiable curiosity, soon convince him he has nothing to fear. Following the shy stage most foals go through a mischievous period when they think it is great sport to bite, kick, and strike. This behavior must not be tolerated. Punishment must be sharp and immediate or these high-spirited pranks will become full-blown vices. It may be cute to see a tiny colt mimicking a fierce stallion, rearing and striking; the behavior will be far from cute when the beast weighs a thousand pounds; it also will be far harder to cure. Nip the tendency in the bud with an instantaneous and well aimed swat. Chances are that he will never try that particular trick again.

When the foal is secure on his feet at three or four days it is time to begin his education in regard to his feet. Pick up each foot daily in the same order giving the same voice command. At first he will be unable to balance for more than a second so no attempt should be made to pick out the foot until he is a little older. Then why start this early? The reason is that if the foal offers resistance the handler is stronger at this stage and can easily overcome any disobedience. Respect for the handler's strength

Shetland pony foal. (Owned by Dr. and Mrs. John L. Zintsmaster, Akron, Ohio)

Foals have no fear of people if handled from birth. (Courtesy "Chronicle of the Horse")

carries over to the time when the youngster outweighs him ten times and is many times more powerful. If you doubt the wisdom of starting foot drill at an early age, try picking up the feet of an unhandled or unmannerly adult horse—or of an unhandled weanling for that matter. By the time the foal is a month or two old he will have each foot ready in rotation and will be able to balance long enough to have his feet picked out. By this time, too, he may need to have his feet leveled with the rasp and will submit to the procedure with a minimum of fuss.

When the foal is a few days old he should be fiitted with a halter. It should be loose enough that all the fingers of the hand can be inserted flat between it and his head to allow ample room for nursing and grazing, but no looser or a foot is likely to be caught as he scratches an ear with a hind hoof. Handle the foal daily while he is still in the stall and later in the field as he follows the mare. Never attempt to drag him. If he resists being led, as all foals do at one time or another, have an assistant place a hand on his rump and encourage him from behind. Working alone, pass a rope over the foal's rump and bring pressure on it, never on the halter shank which would encourage him to pull back. Always use the same voice commands, "Come" and "Whoa," and he will soon learn to associate the word with the act.

The foal must also learn to stand tied. This is best accomplished at an early age when there will be less furor and struggle than when the foal is older and stronger. Use a stout halter which will not break, such as a rope halter. Pass the tie rope through the halter ring. Do not use the trigger snap, as the metal will break if much pressure is put on it. Wrap the rope around a pole or tree. Do not tie the rope as it must be able to be released at once if the foal is cast or becomes tangled. Attach the rope low enough so that there is no danger of the foal's being choked if he falls but high enough so that he will not be likely to get a foot over it. All equipment must be in good condition. Nothing makes a horse a puller or halter-breaker quite as efficiently as his ability to snap a frail halter or rotten rope in his first attempt to get away. Successful behavior tends to be repeated and what is done is not easily undone. Most foals will put up some sort of struggle when first tied; the older the animal, the greater the struggle. Many foals which have always been gently handled will take a step back, feel pressure on the poll, and step forward to relieve the pressure. Some handlers believe in encouraging the foal to pull and struggle until he realizes that resistance is hopeless. We prefer to spread the lesson over several quiet sessions and consider it a triumph if the foal learns to stand tied without ever "sitting down."

When the foal will lead, both following his dam and in the opposite direction, and will stand tied quietly to be brushed over and to have his feet picked out, he is said to be halter broken. This should be accomplished before weaning, preferably long before the colt's size and strength are greatly and obviously superior to his handler's.

No phase of management is more important or more likely to be neglected than care of the growing foal's feet. As stated above, his feet should be handled daily almost from birth so that he will be well-mannered. The leg bones in the growing foal are soft and cartilaginous and are easily deflected from the desirable straight line of growth. If the feet are worn

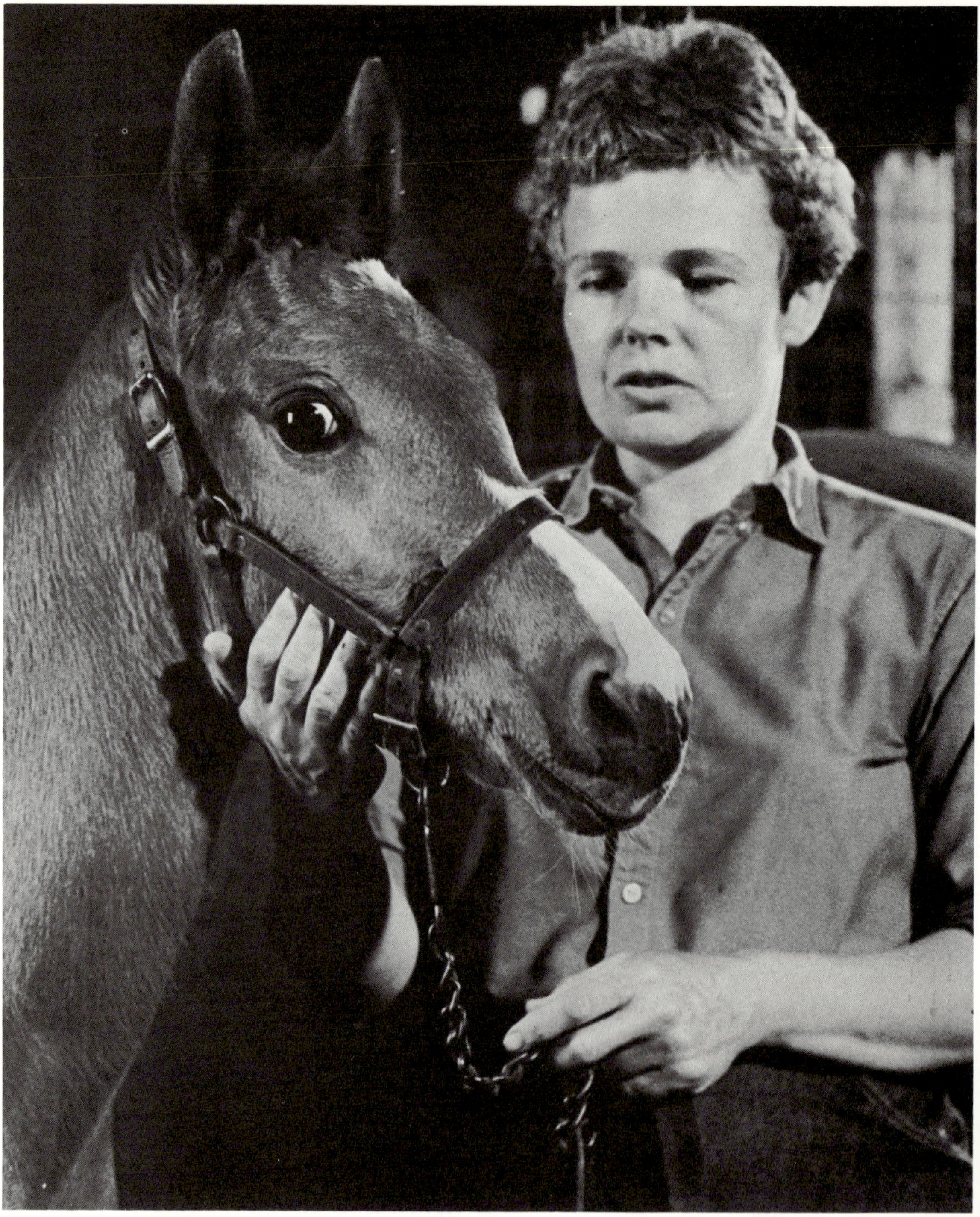

Month-old Morgan filly. The halter should be adjusted to fit snugly, but loose enough to allow you to slip fingers between it and the foal's cheek. (Photo by John Boardman)

down unevenly, if they are allowed to grow too long, the horse will have crooked legs at maturity in spite of the best heredity and the best environment in other respects. From a couple of months of age, or whenever uneven or excessive growth is apparent, the feet should be leveled. In young foals when the horn is still soft a rasp is used; older foals with tougher hooves will require nippers, or hoof parers. Some defects which are present at birth may be mitigated by judicious trimming while the youngster's bones are still plastic; on the other hand, a foal of high potential may be made valueless through neglect. Not only may pasterns and canons be deviated from the normal, excessive length also raises the frog so that it is no longer in contact with the ground; lack of frog pressure allows the foot to become contracted, opening the way for navicular disease before the foal is ever put to any use at all.

Shoeing in some respects is to be considered a necessary evil. It undoubtedly damages the hoof wall. Excessive use of the rasp (other than to smooth the clinches) removes the periople, the varnish-like outer coating of the hoof; the hoof then loses its moisture and resilience. Many horses lightly used for pleasure on non-abrasive surfaces never require shoes. Exercise stimulates the growth of horn so that all but extraordinary wear is replaced. Brittle-footed horses need shoes almost from the beginning regardless of use. The quality of horn is in part determined by heredity, in part by nutritional and metabolic factors such as Vitamin A deficiency. Some horses require shoeing to alter their way of going. Many trainers postpone shoeing as long as possible, waiting until the foot actually shows

Neglected hooves. Note the distorted foot and the oblique angle of the pastern.

Thoroughbred suckling filly. Note the straight hind legs, good bone, and large, well-developed hocks.

Brushing the foal. (Photo by John Boardman)

signs of wear. It is thought that toe grabs contribute to "shin buck" (inflamation and even green stick fracture of the periosteum) in young Thoroughbreds. Other trainers feel that it is better for young horses to get used to shoes early, as they will need them eventually.

Although a foal does not require daily grooming, part of his early training is getting him used to being brushed all over, including the ticklish spots under the belly and inside the hind legs. Most foals very much enjoy being curried, but gentleness must be used around the sensitive face and ears, as rough handling leads to head shy horses. Most foals will reciprocate by trying to curry the handler—sort of "You scratch me and I'll scratch you"; they should be discouraged from becoming too enthusiastic or from nipping.

Foals should be inoculated with tetanus toxoid at six weeks; a second shot is given a month later, followed by a yearly booster shot thereafter to maintain a good level of protection. The antitoxin given at birth gives more immediate but less lasting protection. In areas where viral rhinopneumonitis is endemic (such as central Kentucky), all stock is inoculated intranasally twice yearly in July and October with virus slightly modified by propagation in hamsters; there is a relatively mild respiratory infection followed by temporary immunity. Inoculated horses must be kept isolated from others for three weeks, as the induced virus is transmittible. Once undertaken, the immunization program must be continued, so it should neither be undertaken lightly nor until a positive culture for viral rhinopneumonitis has been obtained. Annual vaccination against equine encephalomyelitis in certain areas (Eastern Seaboard, Gulf Coast, and Far West) is also recommended. Mosquitoes and birds are thought to be vectors in the spread of this highly fatal (90 per cent for Eastern strain) disease. There is a bacterin newly on the market which gives immunity to *Streptococcus equi* that causes strangles.

9
Weaning

WEANING IS A CRITICAL PERIOD IN THE FOAL'S LIFE. IN A NATURAL STATE, foals continue to run with their dams. If the mare is in foal again the milk supply will gradually diminish as pregnancy advances, drying up a couple of months before the next foaling so that the yearling is automatically weaned. If the foal is allowed to run with the mare indefinitely the mare will go down in condition, especially if she is in foal again. She may prefer her older offspring, allowing it to nurse to the detriment and probable death of the newborn. A foal which nurses for a prolonged period will remain excessively dependent on his dam rather than outgrowing his extreme attachment if separated at the usual time.

It is inadvisable to wean before five or six months, although the condition of both mare and foal are more important to consider than an arbitrary date. This is one real advantage which the small breeder has over the large one: he can play it by ear, weaning when it is best for the individual rather than all at one time as farms which wean a hundred or more foals must do—usually the latter part of September, which works to the disadvantage of May foals. Before weaning takes place, the colt should be well grown, in good flesh, and eating six to eight quarts of grain a day. If he seems to be doing poorly, weaning should be postponed until he is on a more secure footing or he will suffer a severe setback when deprived of the mare's milk. He may be encouraged to eat more grain by pouring molasses over it, adding sweet feed to the ration, or even by shutting him away from the mare for brief periods till he develops an appetite.

If the mare is in foal again, if she is a heavy milker and somewhat down in flesh, the foal should not be allowed to run with her beyond six months.

Where has my mother gone?

If he is an exceptionally good doer and she needs to be built up he may be removed as early as four months, although the setback to him will be more marked than if he were older.

Ideally, the preparation for weaning should begin when the foal is a month old, at which time the mare begins to be ridden again. Gradually the period of separation may be lengthened from a few minutes, as when the mare is led out to be serviced at thirty days, to several hours when she is again fit enough for extended work. Many mares, especially those who have had more than one foal, take this separation calmly. Some mares are so frantic that riding them is impossible while they have a foal at foot. All foals are wild with anxiety when first separated from their dams. They must be confined in a stall with solid walls all the way up and with neither crevices to catch a foot nor feed tubs, buckets, mangers, or other projections to cause injury. The foaling stall which follows the same specifications may be used. The foal will be quieter if he is shut up with a companion, another foal or a gentle pony. If separation is repeated regularly, frequently, and for gradually increasing periods of time, the foal will be calm about being left; as he becomes older he may stay in the paddock instead of the barn but always with a baby sitter. By the time he is six months old he will be used to the mare's comings and goings and less likely to panic when the final separation takes place. This is another real advantage which the backyard breeder has over a large farm with many mares which are never ridden and whose only function is to produce.

Another method of weaning, the one most commonly used by those who handle a great many horses, is to leave the foals, one or two to a box stall, and take the mares by truck to a distant part of the farm where they are completely out of hearing. They never are permitted to see each other again. The foals are kept up the first day and then turned out in the familiar field in which they have been running. The one or two least promising foals are turned out first and allowed to run down before turning out two more, and so on, saving the best individuals till last so that they will be the least likely to be injured.

We prefer to place mare and foal in separate but adjoining box stalls. The foal is allowed to nurse three times on the first day, after which it is returned to its stall, twice on the second, and once on the third day, after

Friendly Morgan weanlings. (Courtesy of Barney Caplette, Charlton, Mass.)

which mare and foal are not permitted to run together for at least six weeks at the end of which time the mare should be thoroughly dried up. Gradual weaning makes drying up easier for the mare with less likelihood of a caked udder than in sudden separation. The familiar baby sitter is of the greatest help at weaning whether it is goat, pony, or gentle older horse which is not likely to be stirred up if the foal "runs his head off" or insulted if he attempts to nurse.

During weaning the foal must have water available at all times as his need for water will be far greater than before, due to his being deprived of fluid—in some cases as much as 19 quarts!—in the form of milk. Close attention must be paid to the consistency of his manure as he is likely to become constipated. If he is passing hard, scanty balls of dung proper steps must be taken. As a preventive, feed the weanling a laxative diet containing bran, linseed, and alfalfa.

The foal that is not cleaning up four to eight quarts of grain daily, the exact amount depending on his size, breed, temperament, and rate of growth, will suffer a temporary setback when he is weaned. Goldy was such a heavy milker that Blackacre Sam, her first foal, scarcely ate any grain. At weaning he lost condition rapidly until he developed a taste for oats and sweet feed. At the other extreme Beverly Blue had very little milk when Night Season was a foal; he began to eat grain at three days and at six months could consume as much as a mature horse in active training! He showed no setback at all when weaned.

It is good practice to sprinkle powdered milk over the foal's grain for the first few weeks; this is particularly beneficial for weanlings that do not seem to be doing very well.

It is not uncommon for the foal to develop a cold shortly after weaning. Due to the abrupt decrease in nourishment he is somewhat below par and is tired out and distressed from constant running and calling for his dam. He is more subject to infection from one of the bacteria or viruses which are always present in stable and corral areas when he has been deprived of the antibodies contained in his mother's milk. If there is any significant rise in temperature, severe loss of appetite, persistent cough, discharge of food particles through the nostrils, or thick, purulent nasal discharge, a veterinarian should be consulted.

A few days before weaning the mare's grain and water are sharply reduced to help dry up the milk supply. If her udder becomes painfully full and swollen pressure may be relieved by milking her off a little; but only a little: if she is milked dry she will be stimulated to continue producing milk. In milking several mares, the attendant should wash his hands thoroughly after each to prevent mastitis being spread from infected to healthy mares. Camphorated oil rubbed on the udder hastens drying up. Mares which are exceptional milkers sometimes must be given hormone shots to cut off the supply, especially if they are barren. As a rule, mares that are in foal dry up promptly while barren mares continue to have milk for a year or more in diminishing quantity.

If the mare is drying up without complication or need for further attention she may be turned out, being fed grain as her condition warrants. If she is run down, she must be built up before foaling again. If she is barren care should be taken that she not become too fat, as overly fat barren mares are notoriously hard to get in foal. If there is shelter of some sort the mare may winter out of doors until shortly before she is due to foal again. Old mares should be stabled at night in severe weather. If the mare is believed to be barren, she is caught every other day beginning the second week in February and is brought in to be teased to determine if and when she is in season and to be bred if it is thought to be advisable.

10
Yearlings

YEARLINGS ARE ALLOWED TO GROW AND DEVELOP, SPENDING AS MUCH TIME out of doors as the weather permits. Their feet require periodic attention, both to insure their even, regular growth and to maintain the youngster's good manners. They do not require or even particularly benefit from daily grooming, but should be brushed over from time to time to keep them accustomed to being handled. They are wormed twice a year or more often if their condition warrants it; if they are in small quarters which are more or less continually contaminated, low level treatment is probably preferable. When not on lush pasture they should have all the best mixed hay they will eat, a vitamin supplement if there is any question of soil deficiency where the feed is grown, and grain in accordance with their growth rate. For instance, the Thoroughbred filly Blue Hour, who is developing very rapidly, consumes twice the concentrates that the slower growing Morgan, Debt, or the half-Arab, Grace, are able to eat.

Yearlings intended for the public sales lead quite a different life. They are over-fed, over-confined, protected from sunburn, groomed, and pampered. They are taught to stand just so, in order that they may make the best impression on sale day. They are even shod in some instances to overcome as far as possible the appearance of any leg or foot defect. The colts are not turned out together because of the risk that they might acquire a blemish which would lower the hoped-for sale price by several thousand dollars. Most knowledgeable people buying sale yearlings in overly high condition turn them out for several months to tighten them up and lose some of their excess weight before any further training is attempted.

Long yearling Thoroughbred colt.

Yearling with mane braided to train it to lie on the off side.

Well-grown yearling Thoroughbred filly. (Chris Christenson photo)

Thoroughbreds which are to race at two begin their education at eighteen months. First they are walked while being led with a lightweight boy up. Later they are jogged and by late fall of their yearling year they are being galloped or even breezed!

At this stage we prefer to concentrate on ground work, beginning with longeing. An assistant is necessary for the first few lessons. A special longe (also spelled "lunge" to conform with the pronunciation) cavesson with a ring on the front of the noseband makes the job easier, although it may be done with a halter. A longe tape with a snap and swivel is preferable to an ordinary rope. A buggy whip with a long lash completes the necessary equipment. A small paddock or the center of a riding hall (if you are lucky enough to have one) is an ideal place to begin longeing. A level, confined area with few distractions is the most satisfactory place to work. In the beginning keep the circle small, letting out more loops of tape as the colt gets the feel of the operation and realizes what is expected of him. To work to the left, hold the tape in the left hand, whip in the right, always keeping the horse between the two hands. The assistant walks at the colt's off side, leading him with a shank. The colt is asked to walk on and to stop

Teaching the yearling to longe.

by the voice commands that he learned during halter breaking. The whip is not to hit but to encourage him to walk on and to prevent his coming into the middle of the circle. A surcingle or harness back band and girth fitted with side lines or an overcheck is a convenience in preventing grazing when working in a grassy area; it also gets the colt used to the feel of something on his back. It is not an absolute necessity at this time, however.

Longeing not only supples the colt and makes him more obedient, it also serves as a most valuable asset later in his life, as he may be exercised in a small space when it is not feasible to ride him. Many trainers prefer to take the edge off a horse that has become overly fresh by longeing him for fifteen or twenty minutes before working him under saddle.

When the colt seems to understand what is expected of him and is walking out willingly and calmly, the assistant unsnaps his shank and stands to one side. The radius of the circle should be enlarged gradually. The colt should not be asked to trot until he is thoroughly obedient at a walk. Again start with a small circle, enlarging it gradually and allowing the colt to move faster as he gains confidence and proficiency. It is a mistake to take up the canter too soon; it is an exciting gait and the colt will be likely to play, bucking and kicking up his heels, which can be dangerous for the handler if he has let out a long length of tape.

The same length of time should be spent working to the right as to the left so that the colt will not prefer one hand to the other. Short lessons are better than long ones. As the young horse's attention span is short, he

Trotting on the longe line.

should not be worked until he becomes bored. Each day it is well to go over the preceding lesson before moving on to the next step. One lesson should be thoroughly mastered before the next is attempted.

When the colt is longeing well it is time to introduce him to the long lines. It is important to take great care in putting on the bridle for the first time as the way in which the colt is handled will determine how he will feel about being bridled on future occasions. There are few more annoying or common bad habits in the horse than resisting the bridle; the vice is directly attributable to mishandling. If the colt offers resistance or seems nervous it is better to take the bridle apart, undoing the cheeks and slipping the crown piece and browband gently over the ears. Then insert the bit and refasten the cheeks. Talk to the colt and praise him. The bit may be a rubber covered snaffle, a special colt bit with dangles for him to play with, or a heavy gauge jointed snaffle, never a wire snaffle which would damage his tender mouth. Let the colt stand bitted for fifteen minutes the first day, either with or without side lines attached to the surcingle. If he has sweated he must be cooled out. If he was calm the first day, driving lines may be attached to the bit after he is bridled the second time. Lines with a trigger snap at the end are far easier to manage than those with buckles or hook-in studs. The lines are run through the large rings on the surcingle, the terrets or shaft tugs of a harness back band, or the run up stirrups of a saddle; the purpose of this is to prevent the lines from dragging on the ground where the colt would be sure to step on them.

With an assistant at his head the colt is worked in the same enclosure in which he was longed. The trainer walks behind, encouraging the colt with his voice. He walks far enough behind to avoid being kicked should the colt become frightened. The colt is asked to stop, start, and turn. Ten minutes is enough for the first lesson in driving. The next day, if all is going calmly, the assistant unsnaps his shank. A cooperative colt after a few days' practice in the ring or paddock may be line driven all over the farm. He will enjoy this a great deal and will not become bored or stale as he would if the training continued indefinitely within an enclosure. In this way he will become bridle wise before he is ridden for the first time, without being confused by the unaccustomed weight on his back and the legs touching his side.

CASTRATION

Castration is usually performed between weaning and two years of age. A colt with a short, thick neck or one that behaves in a "studdy" manner is a candidate for early castration, while a slight, weedy, slow to develop type would best be allowed to acquire as much as possible the masculine secondary sex characteristics. Although castration may be done with the greatest ease and safety at a few weeks the result is the imperfect development of the foreparts. Castration delayed long past puberty is more risky and the desired results do not come immediately—*i.e.*, the stallion cut after several years of active service will retain his libido for some time. In aged stallions there is a greater danger of hemorrhage than in younger subjects due to enlargement of the spermatic blood vessels. As a rule, Thoroughbreds are allowed to start as colts so that some idea of their potential may be determined. Some colts do not train well as entires, however, being too interested in other things to concentrate on racing. A poor doer will often put on weight after being altered.

All colts should be critically examined to determine whether they are outstanding enough to be considered as future herd sires. Castration is the horse industry's most efficient means of quality control. One of the very real hazards of a breed's rise to popularity is the failure to castrate mediocre males. Almost anyone who has bought an expensive purebred colt entertains the hope of earning back at least part of the purchase price in stud fees, regardless of the animal's shortcomings. This can only make for a rapid deterioration of the breed.

Although castration at any age may be performed by a skilled practitioner without undue hazard, the operation should not be done in fly season when the fresh blood will attract swarms of insects to torment and promote infection in the new gelding. Castration is followed by an injection of tetanus antitoxin. The gelding is given the freedom of a paddock or, if stabled, is jogged twice a day for twenty minutes to promote drainage. If the incisions close and the scrotum fills with fluid, the wounds must be reopened. Failure to secure proper drainage results in ascending infection and possible death.

11
Two-Year-Olds

IF THE YOUNGSTER IS TO BE TAUGHT TO DRIVE AT TWO, THE VARIOUS PARTS OF the harness should be introduced gradually. He probably will be insulted at having a crupper put under his tail, especially in cold weather. Tact is indicated here! He may be line driven in crupper and britching with the traces dragging to get him used to the straps touching his thighs before he is asked to pull anything. It is a good idea to have him drag some poles around before putting him between the shafts.

He is first hitched to a breaking cart which must be sturdy enough to stand up under hard treatment and not break up if the colt should become frightened and lash out with his heels. An extension at the rear of the cart permits the driver to step on and off with the greatest safety. A kicking strap which goes over the colt's rump discourages kicking.

All harness should be extra heavy duty and in good repair. Any weak or worn spots must be replaced before the colt is hitched, for if he should become frightened and break loose he will think he can get loose whenever he pleases and much valuable groundwork will be undone.

When the colt is first hitched he is led by an assistant as in the previous stages of his training. The driver walks behind the cart, not taking his seat until the colt is working calmly. As before, practice stopping, walking on, and turning. The colt must become used to the pressure of the shafts while turning and to the weight of the vehicle itself pushing on the britching when stopping and against the breast collar when moving. He will need to be reassured that the cart will not run up on him. A blind bridle prevents his seeing it following him and becoming frightened. He should not be asked to trot or to pull up or down hills until he is working dependably on

Driving in long lines.

Two-year-old Morgan colt hitched to the breaking cart. (Courtesy of Barney Caplette, Charlton, Mass.)

level ground at a walk. Begin by jogging, gradually increasing the speed of the trot. He should never be allowed to "break" or to canter or gallop. He must be pulled back to a trot immediately. To allow him to gallop in harness is not only bad manners but invites a runaway.

An overcheck prevents grazing and resisting by getting the head down. The overcheck is attached to a separate bit. The check rein should not be tight, holding the head in an elevated position as is seen in harness racing or in fine harness show classes. The colt cannot pull, especially up hills, unless he has a relatively free head.

What purpose is there in teaching a horse intended for riding to drive? If he is being trained at an early age, driving permits him to put on muscle without having to bear any weight. A horse which is thoroughly broken to drive will be bridle wise and in hard condition before he is ever ridden, thus extending his useful life as well as making his training easier.

Prolonged driving, however, may make the horse reluctant to canter because he has been punished for breaking. He may be an overly cautious trail horse, as he is in the habit of making allowances for a vehicle in turning corners. Goldy, who came to us at the age of eight after a career as a fine harness show horse, pulled her phantom buggy behind her for a considerable time. Even at liberty she still may be seen doing an extended trot while the rest of the brood mare band is galloping.

The colt should not be ridden until he is fully two years old and then only by a lightweight rider for very brief periods. The amount of work he is asked to do should be increased gradually. To require very demanding work such as racing or jumping before four years is damaging to the immature bones of the legs which are incompletely ossified. The result is unsoundness and limitation of the animal's useful life.

If the colt has been handled from birth and if each new step in his training has been introduced only after careful preparation actual riding should be merely another in a series of steps. Begin by putting part of your weight on your hands, pushing down where the saddle goes. Sit on the colt while he is lying in his stall. When he is standing in his stall, lie across his back. Swing a leg across with caution. If the youngster has confidence in you and you are old friends he will allow you to sit on him while he munches hay with no more protest than to nip playfully at your toe. Instinct, however, will tell him to buck, which is how his wild ancestors handled hungry predators that attacked from above. If he is a high-spirited colt he may try to hump up his back the first time he is mounted out of doors, where he feels less secure than in his stall. Therefore, as in the previous stages of training, an assistant, or even two if he is a large, strong youngster, lead him around for the first few minutes until he is quite calm.

Thoroughbred colt and lightweight rider.

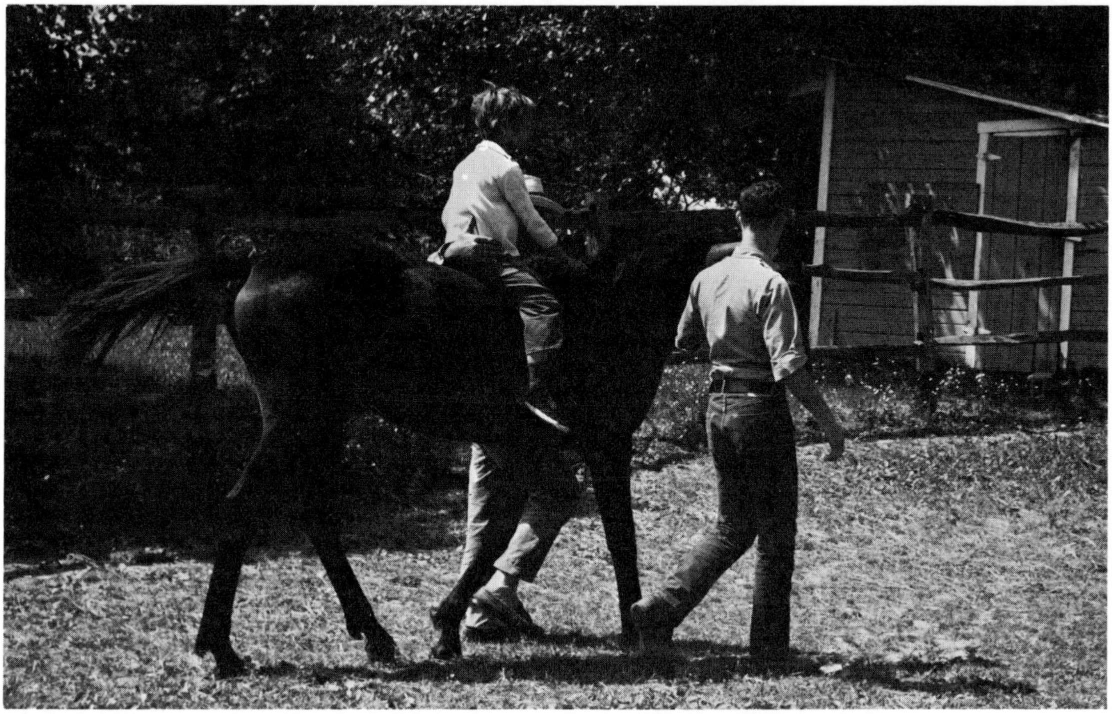

A lightweight rider up for the first time while the colt is being led.

Walk, stop, walk on, and turn the first day, praising and petting the colt when he obeys. He may even be fed a treat while the rider is still mounted so that the colt will associate being ridden with a pleasant experience. At first he will be awkward, heavy on the forehand, and likely to stumble until he learns how to readjust his balance to the unaccustomed weight. For this reason a long time should be spent in walking, first on the level and then up and down hills of increasing steepness, before working at the other gaits under saddle. Begin with a jog gradually working up to a trot. The colt must be fairly reliable before a canter is attempted, as there is something about this gait which brings out the aborigine in even the most sedate horses. Speed should be increased very gradually or the colt is likely to "come apart" or break into bucking or other silliness. Collection at the canter will come gradually as he becomes more supple and balanced. Working over cavaletti, first at a walk, then at a trot, and finally at a canter, adds to his agility, balance, and rhythm. Start with a single pole laid on the ground, adding additional cavaletti as his competence increases.

The colt's back may become sore early in his training, not because the equipment is ill-fitting but because his young skin is still tender. Later it will become toughened by riding. Apply ice to any abrasions. An astringent lotion is helpful in reducing swellings. Many old-time trainers advise the application of salt water to hasten the toughening process or bacon grease to sooth sore spots. The skin at the corner of the mouth may be rubbed by the bit. A pair of leather or rubber discs designed to prevent chafing may be inserted between the snaffle ring and the youngster's lips.

Trailer-breaking is part of the youngster's basic education and it is the responsibility of his breeder to school him in this respect. As in most other instances, the colt's insatiable curiosity works to the trainer's advantage. A trailer, braced, padded, and with the center partition removed, is placed in the paddock. If grain is put in the trailer with an enticing trail leading up the loading ramp the colt, in all probability, will teach himself to load. Night Season was trailer-broken in this way. He used the trailer as a shelter from the wind and made a game out of running in and out. When he had to be shipped to Ohio State University for treatment of an injury, he loaded without incident, although a year's time had elapsed since his trailer-breaking.

Most horses will enter a van far more willingly than a trailer. Reluctant horses should be patted and encouraged and allowed plenty of time. Usually a rope pressed gently across the hind quarters just above the hocks is successful in loading balky horses.

Trailer breaking. The colt's insatiable curiosity plus oats equals walking up the ramp unassisted. (Photo by Robert A. Boyajian)

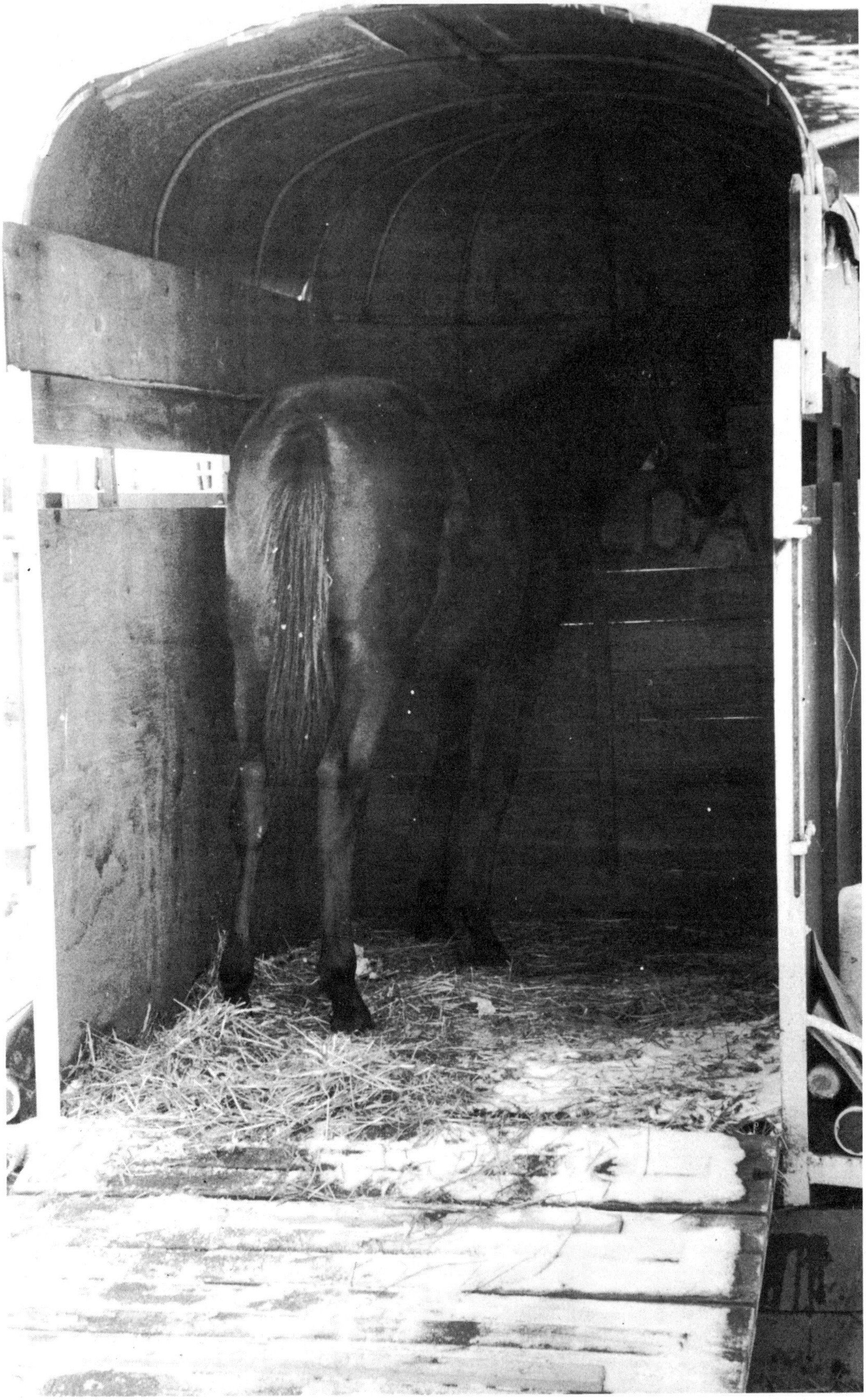

Trailer breaking. The center partition has been removed for safety. (Photo by Robert A Boyajian)

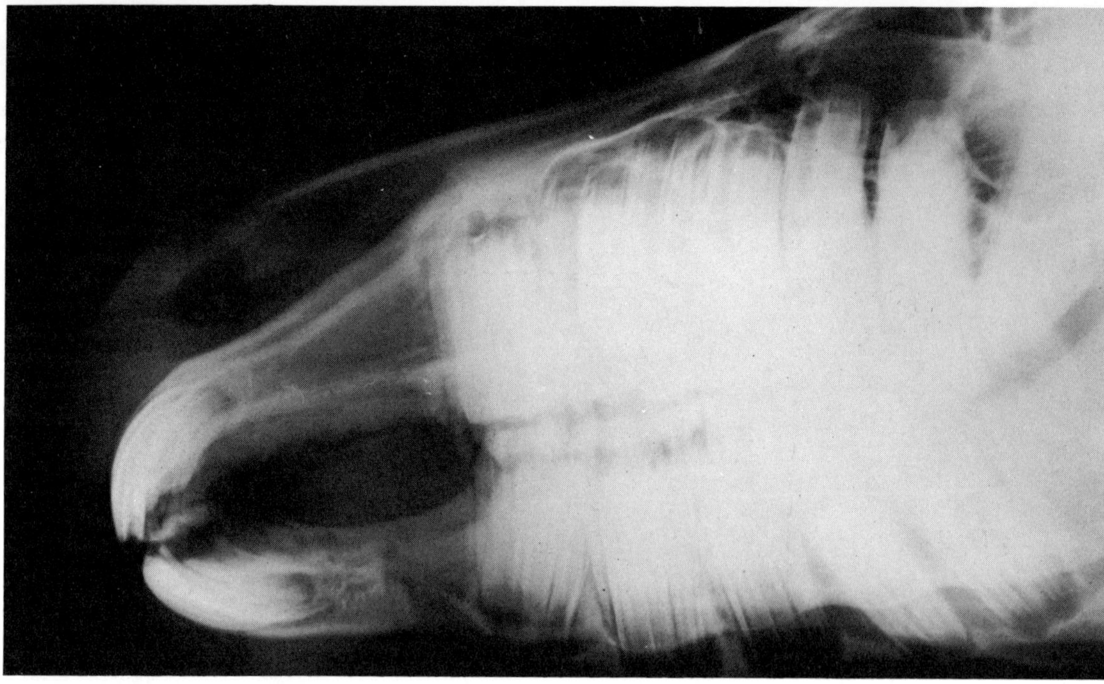

Caps on teeth of young horse. Remains of deciduous molars remain on top of erupting permanent teeth, causing bulges in lower jaw and difficulty in chewing. (Courtesy of Ohio State University)

TEETH

The average time of eruption of teeth is as follows:

Teeth		Eruption
A. *Deciduous*:		
1st incisor	(Di1)	Birth or first week.
2nd "	(Di2)	4–6 weeks.
3rd "	(Di3)	6–9 months.
Canine	(Dc)	[*]
1st premolar	(Dp2)	Birth or first two weeks
2nd "	(Dp3)	Birth or first two weeks
3rd "	(Dp4)	Birth or first two weeks
B. *Permanent*:		
1st incisor	(I1)	2½ years.
2nd "	(I2)	3½ years.
3rd "	(I3)	4½ years.
Canine	(C)	4–5 years.
1st premolar (or wolftooth)	(P1)	5–6 months
2nd "	(P2)	2½ years
3rd "	(P3)	3 years
4th "	(P4)	4 years
1st molar	(M1)	9–12 months
2nd "	(M2)	2 years
3rd "	(M3)	3½–4 years

(The periods given for P3 and 4 refer to the upper teeth; the lower ones may erupt about six months earlier.)[1]

[*] The deciduous canines are vestigial and do not erupt.

[1] Septimus Sirron, S.B., V.S., D.V.Sc. *et al.*, *The Anatomy of Domestic Animals* (Philadelphia: W. B. Saunders Company, 1938).

As the permanent teeth erupt the roots of the deciduous teeth are destroyed and the teeth drop out. Fairly often, however, the baby molars rather than being shed become stuck firmly to the crown of the emerging new teeth, interferring with the animal's chewing and causing lumpy swellings along the edge of the lower jaw. Unchewed, undigested grain appears in the manure. Not infrequently if the condition persists the youngster loses flesh. In such cases the caps should be removed by a veterinarian. The lumps on the jaw will have disappeared by the time the horse is four or five years old.

12
Injuries

WITH YOUNG ANIMALS RISK OF INJURY IS AN EVER-PRESENT CONCERN. FENCES, feed tubs, hay racks, watering troughs, trees, and any other immovable object constitute a menace to the foolish youngster acting like an irresistable force. Something has to give and usually it is tender young horse flesh that is damaged.

Foresight can prevent a high percentage of injuries. Sharp corners and projections should be avoided. The angles of fences in paddocks, corrals, and pastures should be rounded rather than square to prevent a horse's being trapped by an aggressive companion. Use of flimsy materials is an invitation to disaster. Matchstick fencing and stall partitions break easily and splinter into lethal snags and spears that can quite literally cut a horse to pieces. Young stock should be segregated according to age and size to prevent older, stronger animals from bullying younger, weaker foals, stealing their feed, and running them into the fences. Keeping a large number of horses in a small area creates a dangerous situation. Danger goes up in proportion to the number enclosed.

Opinion varies as to the amount of confining necessary to protect young stock. Growing horses turned out together inevitably acquire minor injuries some of which result in permanent blemishes. The risk is more than offset by the toughening process, for foals, if they are to grow to be sound, useful adult horses rather than mere hot-house plants, need daily outdoor exercise. Some farms confine their young stock when there is danger of their falling on ice or other slippery footing. On the other hand, the youngsters' exhuberant foolishness on being released after several days in the barn is probably just as dangerous as the conditions which brought on confining them. In turning a horse out after long confinement, first put him in a small, stout enclosure rather than a large field where he could reach such a high speed that he would be unable to turn or stop for a fence.

Even with the greatest foresight and most careful management injuries are bound to occur where young horses are gathered together in any number.

Minor cuts may be treated by washing with mild soap and water and applying an antiseptic. Wounds below the knee and hock should be bandaged to prevent swelling, to reduce unsightly scarring, and to discourage exuberant granulation (proud flesh). Some practitioners feel that the formation of a hard scab delays healing and advise the application of a fish oil and zinc oxide ointment.

Major wounds require prompt veterinary attention if they are to be treated successfully. In most instances bleeding may be controlled by pressure while waiting for professional help. In cold weather keep the animal warm. Only fresh wounds can be sutured successfully and those in areas where there is not too great tension on the sutures. It is usually necessary to partially immobilize the animal so that the healing process may proceed undisturbed.

Wounds heal by first intention when the cut is cleanly incised, when there is relatively little bruising, and when the raw edges may successfully be held in apposition, either by sutures or bandaging, until a fibrous band of tissue has formed to keep the parts together permanently.

Wounds which do not heal by first intention heal by granulation or the formation of new tissue, composed of capillaries and fibroblasts, which fills the gap between the cut or torn edges of the wound. Eventually fibrous tissue replaces the vascular, making a tough insensitive inelastic scar. In time the scar shrinks. As granulation tissue has some of the characteristics of neoplastic tissue it tends to proliferate wildly, causing the condition known as proud flesh and resulting in unsightly scars extending beyond the original limit of the wound. Exuberant granulation is most common distal to the tarsus and carpus and is discouraged by pressure bandaging.

Puncture wounds are likely to be followed by tetanus if the horse has not been protected against it. If there is doubt as to how recently the horse has had a tetanus toxoid shot he must have a tetanus antitoxin injection at the time of the injury.

Over-use of antiseptics in treating wounds should be avoided although there is a strong desire to "do something." The application of "healing" powders and lotions irritates the tissues and makes the task of regeneration more difficult.

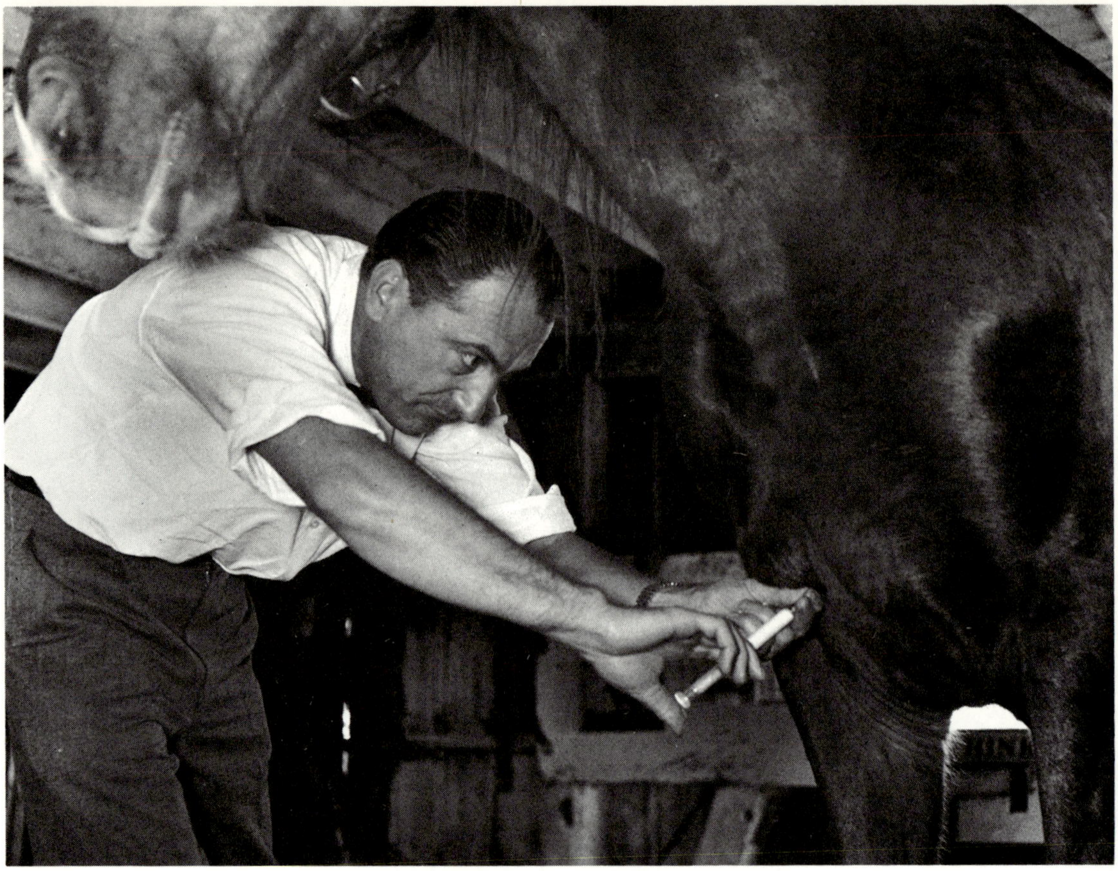

Horse receiving a shot of antibiotic. (Photo by Robert A. Boyajian)

Nearly every horseman has some noxious mixture which has been used for generations and is a fixture in his medicine chest. Actually, the fact that horse's wounds heal so well despite the use of drugs demonstrates the considerable reserve of healing power inherent in the beasts.[1]

The horse has great recuperative powers and often the most appalling wounds heal with little residual scarring or loss of function. Accident-prone Nicholls W. lay down too close to the broken off stump of a young pear tree (which had no business being in his paddock) and on arising inflicted a deep laceration and avulsion to his semitendinosis and biceps femoralis muscles. There was a gaping hole in his thigh into which a man could have put his fist. The wound rapidly granulated forming a hard dark scar that shrank to a thin line now entirely covered by the surrounding hair. Only a careful search discloses the blemish.

Fractures are fairly common. In the past it was axiomatic that if a horse broke a leg he was promptly shot. With improved surgical and orthopedic techniques it is now possible to save approximately 90 per cent of fracture

[1] J. W. Britton, D.V.M., "Wound Treatment," *Equine Medicine and Surgery* (Santa Barbara: American Veterinary Publications, Inc., 1963) p. 702.

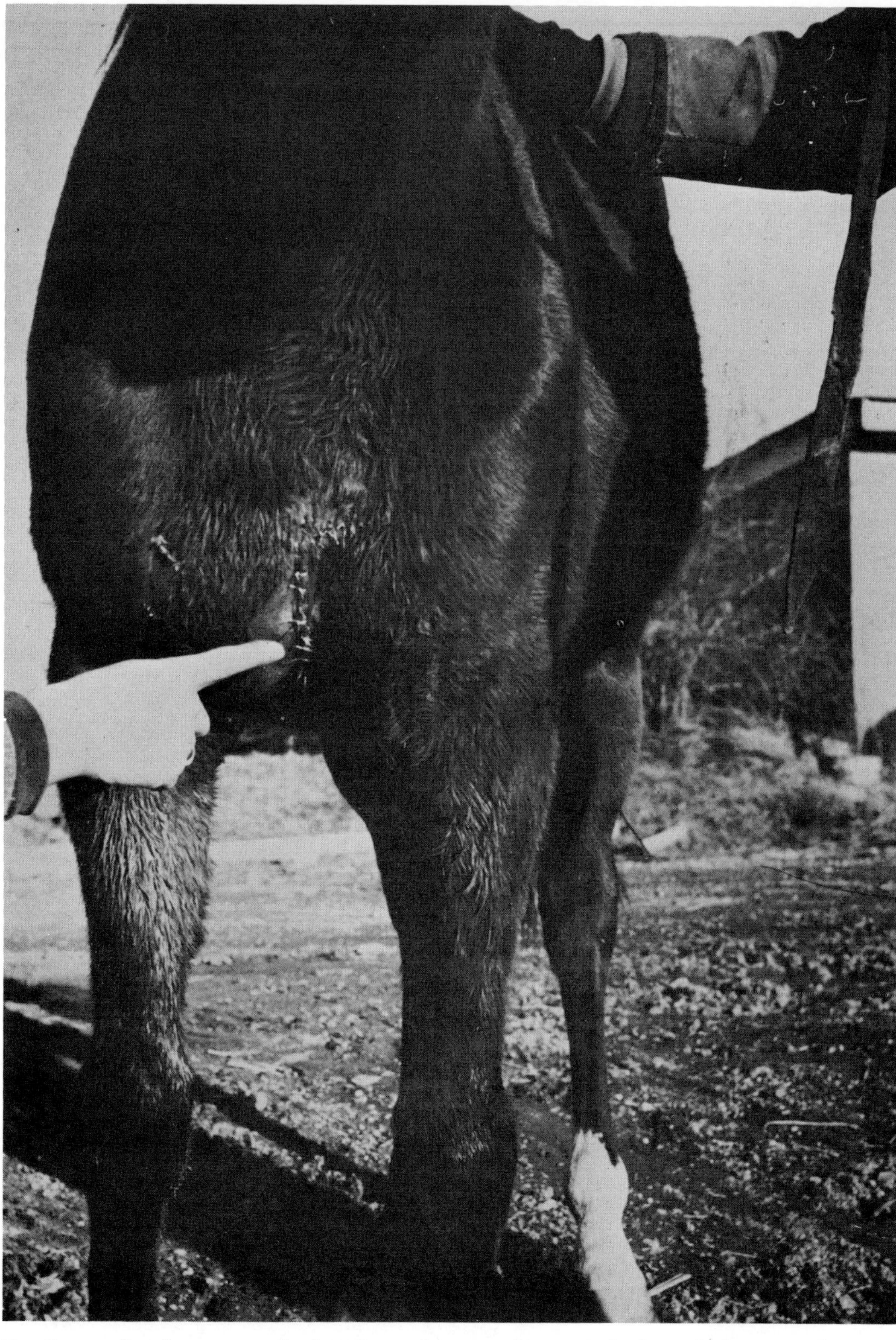

Dot Sena suffered an extremely deep laceration extending entirely through the pectoral muscles to the sternum. The wound is held together with both deep and skin sutures. At the time this picture was taken, about two weeks after the accident, the swelling had partially subsided.

The superficial sutures have been removed (the deep sutures of gut are absorbed), and the swelling has disappeared almost completely. New hair is growing in to replace the hair that fell out in the injured area.

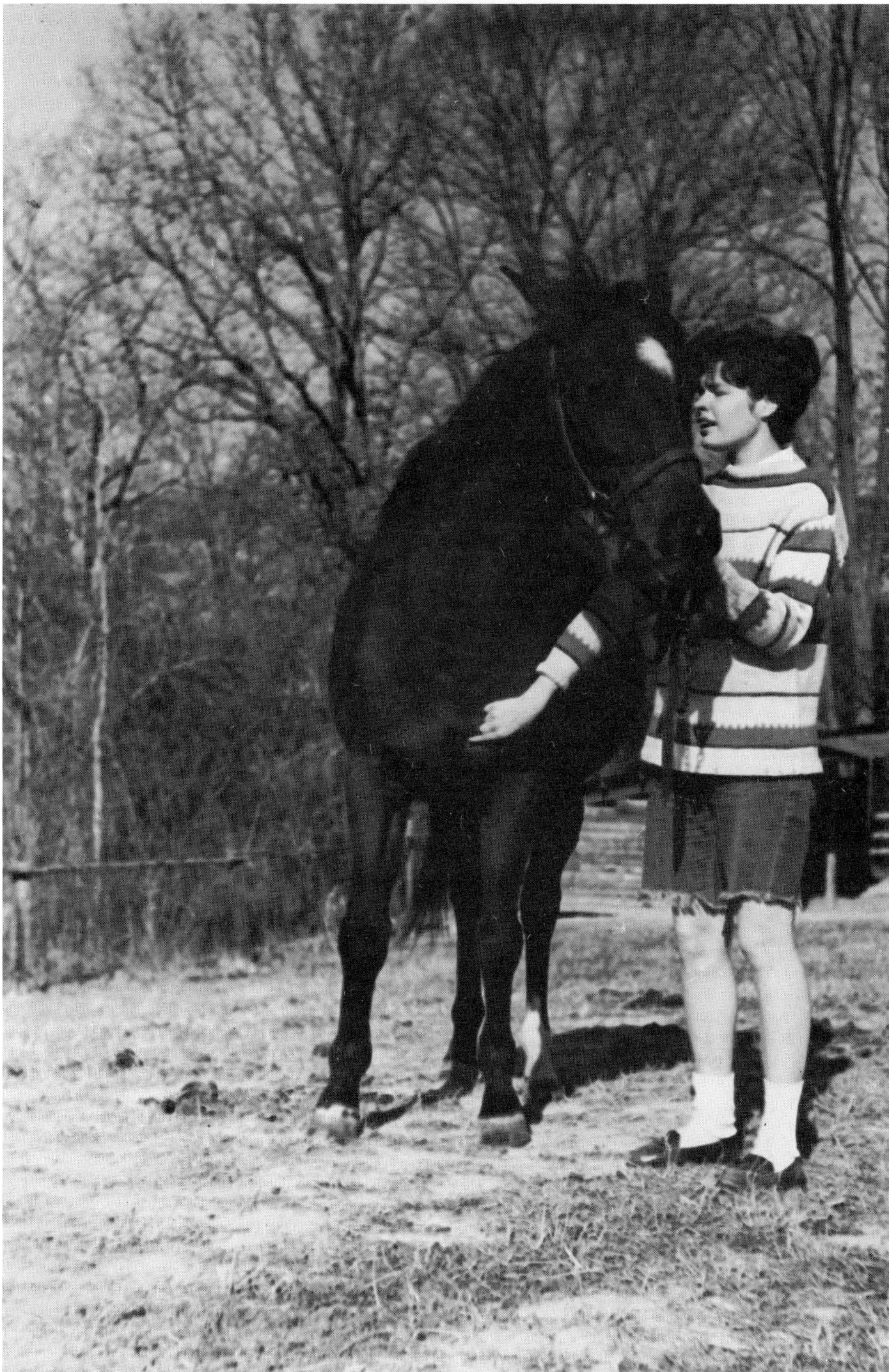

The wound is marked by a thin scar which the horse's winter coat almost completely conceals. Ten months elapsed between first and last pictures.

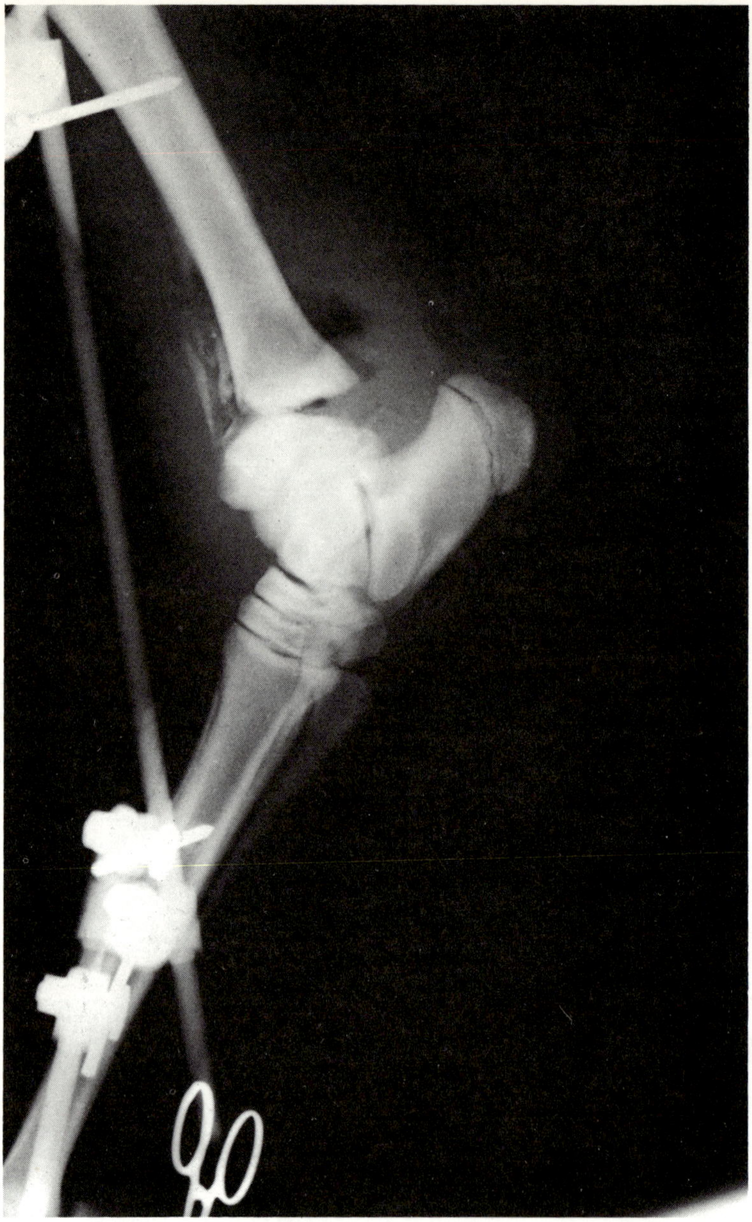

Fracture of tarsus (hock) immobilized by rod attached to pins in tibia and metatarsal. (Courtesy of Ohio State University)

cases which receive prompt and proper treatment. This is of the greatest importance in potential breeding stock; it is often not practical, humane, or economically feasible to try to save geldings if there is likely to be a great deal of residual impairment of gait. Successful reduction of the femur and humerus is difficult because of the force exerted by the strong pull of shoulder and thigh muscles. Because the fragments override each other there is inevitably a degree of resulting deformity and lameness due to shortening of the limb. In 1951 Your Host, stakes winner of thirteen races

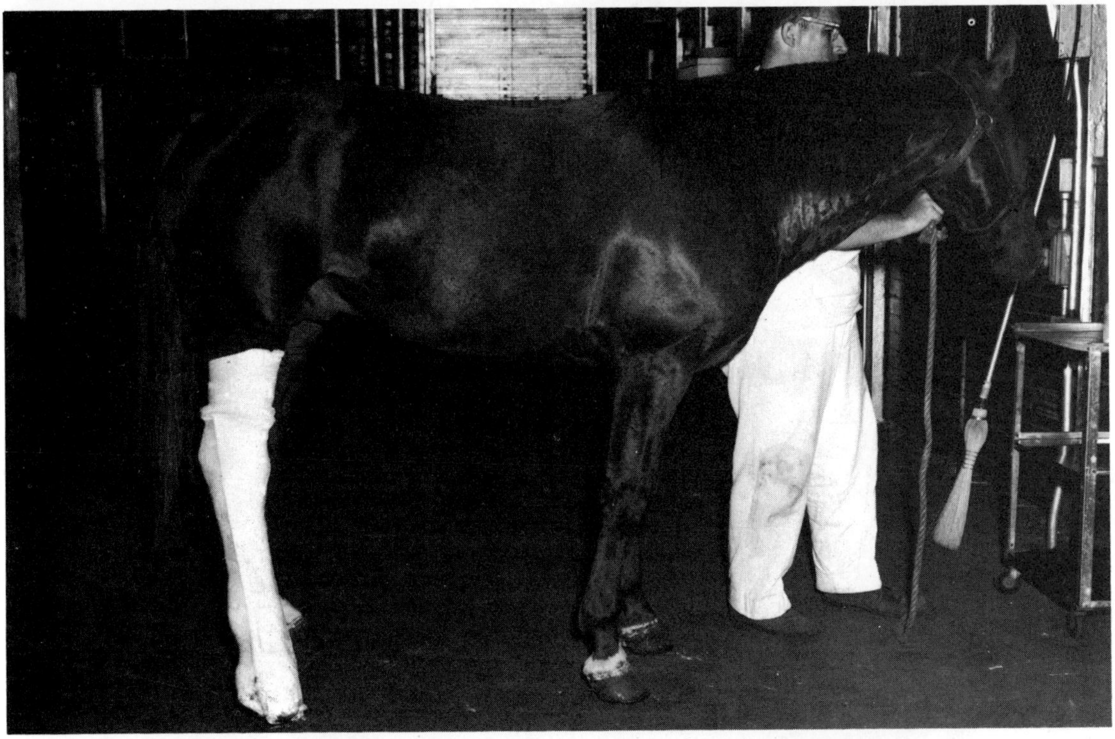

Colt in a walking cast. This major contribution to the successful treatment of equine fractures was developed at Ohio State University under the direction of L. E. Johnson, D.V.M. (Courtesy of Ohio State University)

and $384,795, fractured his ulna in a race at Santa Anita. Lloyd's of London paid the claim in full to the horse's owner, but instead of destroying him had the injury treated and sold him to a syndicate. Although permanently crippled, he sired all-time money-earner Kelso, Handshake, Windy Sands, Three M.R., and other notable runners before being put down in 1962 due to a dislocated stifle which proved impossible to treat in light of the earlier injury.

TYPES OF INJURIES

Wounds:

Lacerations are tears with jagged, uneven edges. Unless a major vessel is involved they seldom bleed profusely. They are caused by barbed wire, wood fences, and stall partitions.

Incised wounds are cleanly cut, as by broken glass or sharp protrusions of metal.

Avulsions result from the tearing away of tissue and are caused by barbed wire, sharp fence posts which protrude above the top rail, broken tree limbs, automobile bumper guards, etc.

Punctures are relatively deep with a small opening. They are caused by

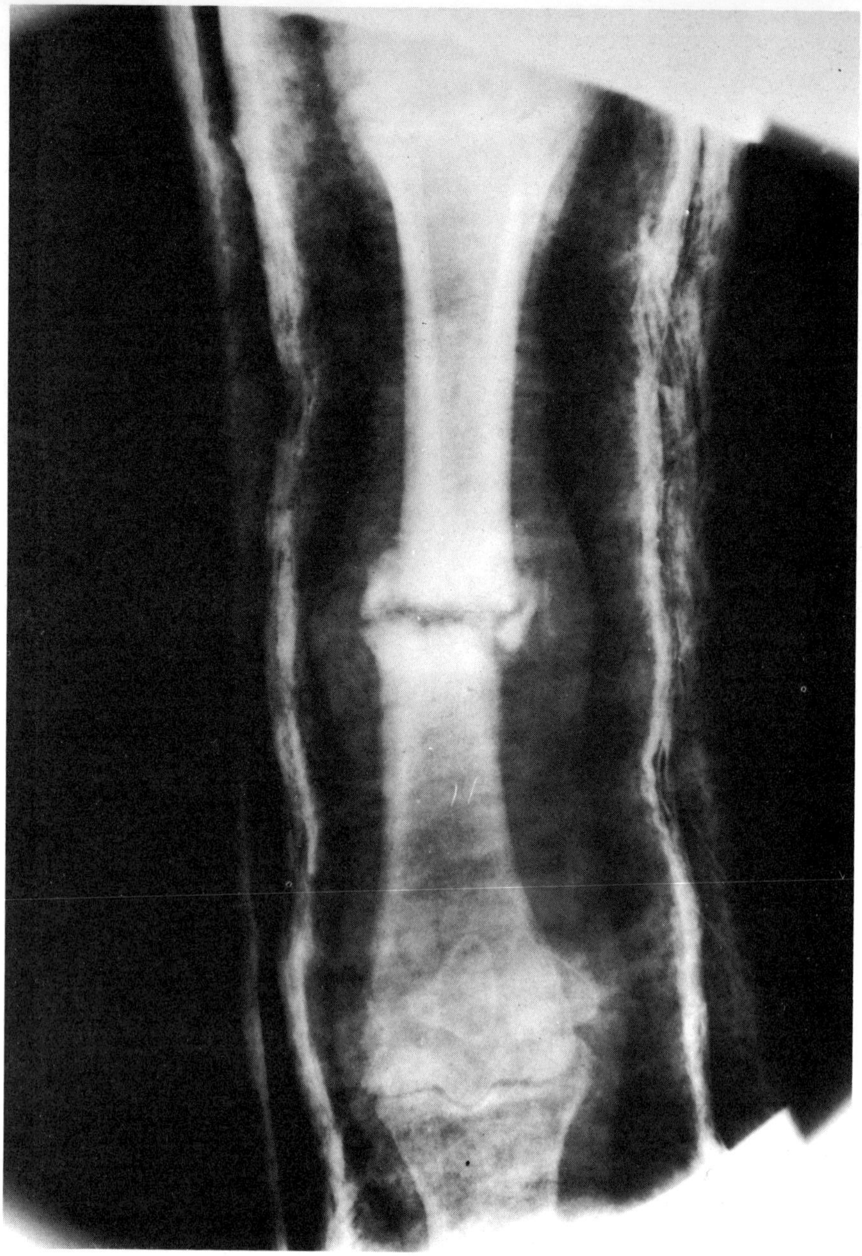

Fracture that failed to unite. (Courtesy of Ohio State University)

nails, splinters, and thorns. They are likely to become infected due to poor drainage. Tetanus is also probable in unprotected horses.

Contusions are bruises or injuries in which the skin is not broken. They are caused by kicks, blows, falls, and running into fixed objects.

Hematomas are localized bleedings which soon clot and become encapsulated by connective tissue. They are caused by kicks, blows, and running into fixed objects.

Abrasions are a scraping away of the hair and surface skin caused by

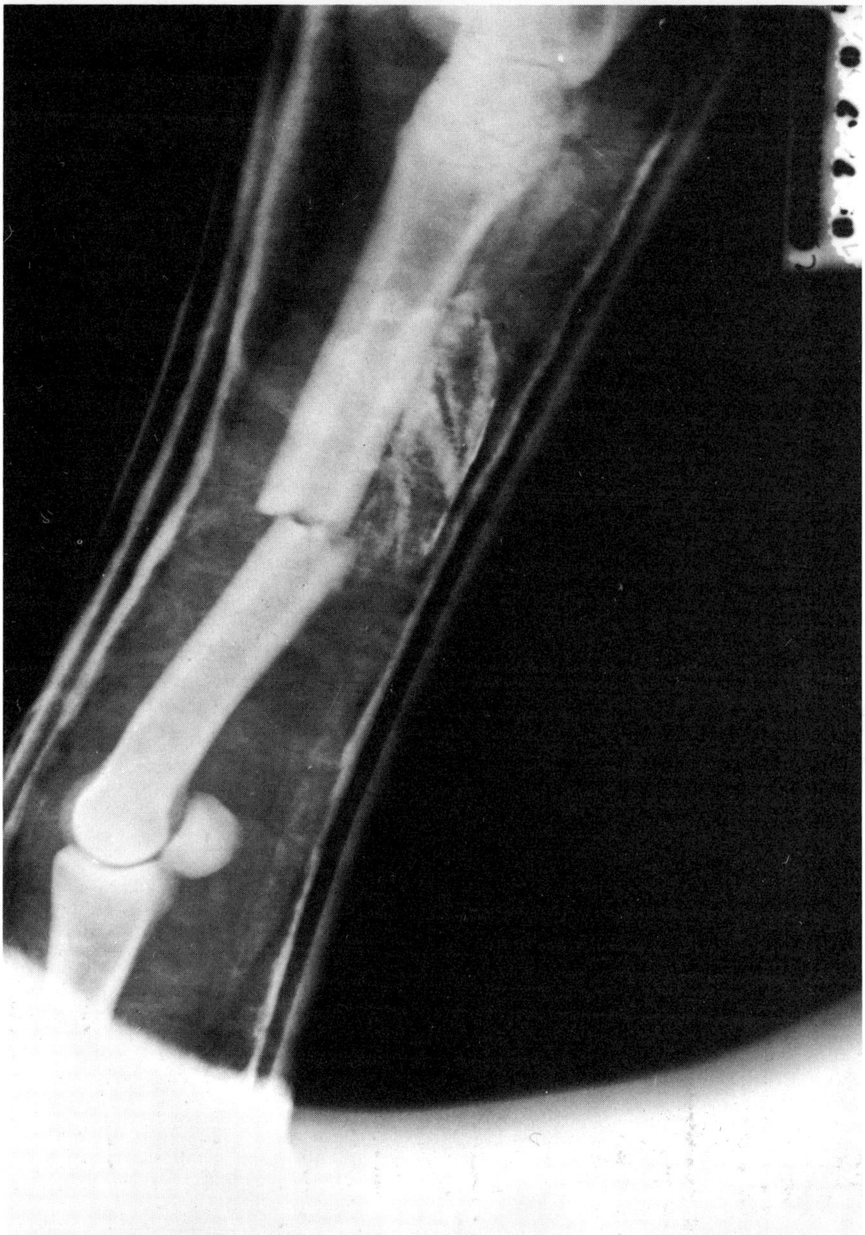

Complete fracture of large metacarpal. (Courtesy of Ohio State University)

friction resulting from falls, rope burns, scraping against loading ramp of van or trailer.

Fractures:

Simple. One in which there is no communication with the outside, or the skin surface.

Compound. Portion of the fracture in contact with the external surface of the horse's body.

Comminuted. A splintering and fragmentation of the bone. The prognosis is poor except in young animals.

Compression. A crushing in which one surface of a bone is driven toward another surface; commonly found in vertibrae; caused by blows, violent falls, and highway accidents.

Chip. Small piece broken away from main body of the bone; due to trauma or stress.

Depressed. When part of the skull is depressed below normal level; due to kicks, blows, low hanging branches and projections, rearing in van or trailer.

Epiphyseal separation. The breaking off of an epiphysis from the shaft of the bone; due to stress or trauma in immature animal.

Complete. Continuity of entire bone is destroyed.

Incomplete. Not extending through the entire bone.

Pathological. Due to diseased condition of bone rather than external violence.

Greenstick. Incomplete fracture of long bone of immature horse; the bone is bent but is splintered only on convex side. Shinbuck is thought to be a greenstick fracture of the metacarpal (canon) bone.

13
Heredity

FROM THE TIME THAT A MATING IS PLANNED, CERTAINLY FROM THE TIME that the mare is believed safely in foal, there is speculation as to what the foal will be like, both at birth and at maturity.

His color, marks, temperament, and potential for growth and conformation are fixed at the moment of conception. Each parent furnishes one half of a set of paired chromosomes which he in turn received from his parents and so on in a living chain back to the beginning of all life. Before conception the gametes (male: sperm; female: ovum) undergo chromosome reduction. The cell divides by meiosis; the paired chromosomes line up in opposite halves of the cell which then splits. Thus each parent passes on only half of its sixty-six chromosomes.

The gamete after meiosis is called a haploid. The haploid combines with the haploid of the other parent, the number of chromosomes within a species thus remaining constant. The chromosomes contain countless genes which are the carriers of heredity, determining specific characteristics such as blood type, eye color, quantity and quality of hair, size and shape of ears, size and position of eyes, intelligence, etc.

The assortment of chromosomes within the haploid is determined by the laws of chance. There are two to the thirty-second power combinations possible or two to the thirty-third power (billions!) in the fertilized ovum. In theory, full brothers and sisters would tend to average 50% of the same genetic material while in fact they might have from 0% to 100% of the same inheritance, which accounts for the often striking dissimilarities between full siblings. Among the offspring of closely related sire and dam there is less potential for difference.

Organisms are said to be heterozygous or homozygous. Homozygous horses are those which are the result of matings which draw from a limited gene pool, as is the case in most registered stock in which there has been a certain amount of inbreeding, or the breeding of related individuals. In a state of nature inbreeding takes place when a group is isolated geographically, and a very definite, uniform type develops. Line breeding, a less intense form of inbreeding, is practiced to obtain a high percentage of the "blood" of some admired ancestor. The more similar the pedigrees of sire and dam, the more ancestors they have in common, the more accurately it is possible to predict the characteristics of the foal.

Heterozygous horses are crossbred, "mongrels," the result of mating unrelated animals. While often outstanding individuals in their own right, their ability to reproduce themselves is uncertain due to their varied background. They do not breed true. The first generation of a cross is likely to be larger, stronger, and healthier than either parent. This is the famous hybrid vigor. "Nicks" are fortuitous crossings of strains within a breed which produce a superior horse. Man o' War was the result of the Hastings-*Rock Sand cross (by Fair Play by Hastings, out of Mahuba by *Rock Sand). The *Khaled cross on *Beau Peu mares produced the notable runners, Swaps, Candy Spots (through his dam Candy Dish), and El Drag. It is thought that stallions imported to the North American continent do better in the stud than homebreds because they introduce new genetic material which revitalizes the local strain.

When two different genes for a characteristic meet which takes precedence? Some genes are dominant or manifest whenever they exist. Others are recessive, needing two of a kind to be evident. They may be carried for generations, being masked by a dominant, until, paired with another like gene, they crop up unexpectedly. Because a dominant must show itself if it exists, it is relatively easy to eliminate completely from a breed, as pinto patterning has been weeded out of every light horse registry.

Most defects are recessive. The risk of two individuals both carrying, masked, the same undesirable trait is greatly increased if they are closely related. Therefore very close inbreeding—the mating of brother and sister, mother and son, father and daughter—is not usually practiced with horses, there being little market for the undesirable results of experimentation. Inbreeding *per se* is not undesirable. The Arabs believed that a strong line prospered rather than declined when closely mated. Probably the only undesirable effect, other than the greater risk of a hidden defective recessive coming to light, is a reduction in size.

A recessive characteristic may be fixed readily. If one wished to breed only chestnut horses, bays and other colors dominant to chestnut could

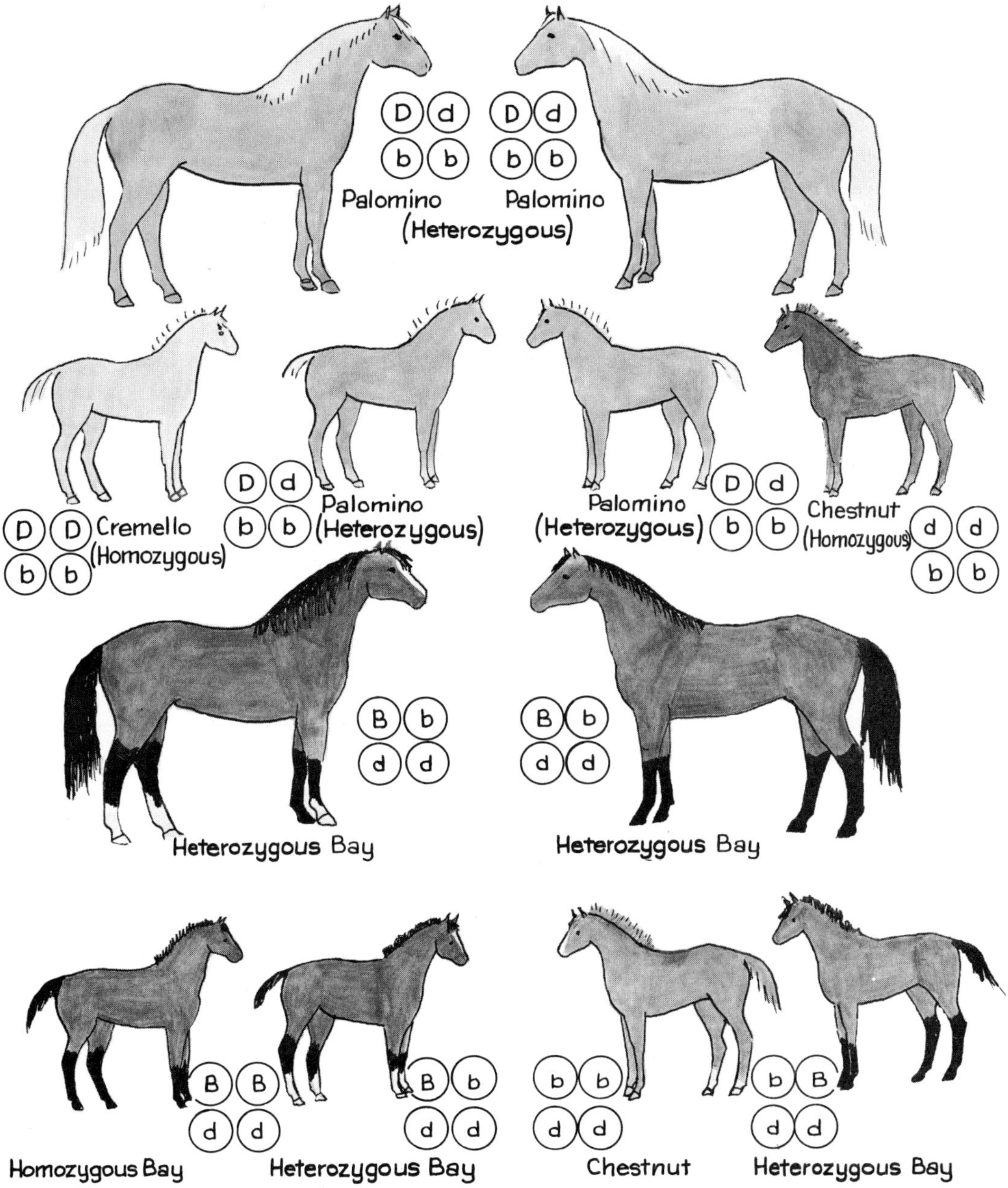

Possible coat color of foals of heterozygous parents. The homozygous cremello, chestnuts, and bay could only produce foals like themselves if mated with other homozygous animals of the same color.

effectively be eliminated in a generation. The Suffolk breed of draft horses are all chestnut. A dominant trait, however, cannot be fixed because it is impossible to tell from the animal's appearance (phenotype) whether he carries a masked recessive gene for the trait in question. To the eye a homozygous bay is indistinguishable from a heterozygous bay. Genetic composition may be analyzed by a back cross, or crossing with a recessive.

To determine the genotype, dominant bay is crossed with recessive chestnut; if the bay is homozygous all the offspring will be bay, while if he is heterozygous, half of the foals will be chestnut. St. Simon is an example of a homozygous bay, incapable of siring chestnut foals. All of his 436 offspring were bay or brown with the lone exception of the gray colt Postumus, his very last foal.

Certain individuals carry a larger than usual number of dominant traits and are said to be prepotent. The most fully prepotent individuals are also homozygous. For a stallion to be a success in the stud he must be prepotent in regard to his best characteristics or he will get good foals only on outstanding mares.

Some characteristics are sex-linked. A mare passes a trait to her sons but not to her daughters; it, in turn, is inherited by the sons' fillies but is absent in colts sired by them. Thus some stallions are known as outstanding brood mare sires while getting only mediocre male offspring.

In the male there are thirty-two pairs of homologous (like) chromosomes plus one pair of unlike members, the larger of which is known as X and the smaller as Y, which is often described as incomplete. In females the pair is identical: two X chromosomes, thirty-three homologous pairs. During meiosis in the male gamete the sperm receives an X or a Y according to the laws of chance. An ovum fertilized by a Y will produce a colt, an X, a filly. Although X- and Y-carrying spermatazoa are equally divided in number, the Y's are more motile and more males are conceived. Two X chromosomes are more effective in promoting resilience and longevity to help offset the demands made on the female organism during pregnancy, foaling, and lactation. In this sense the male is, in effect, an imperfect female. The male has a more primitive genito-urinary tract, with the urethra serving as a passage for both procreative fluid (semen) and excretory fluid (urine).

Sex-linked recessive characteristics are inherited in this manner. The X in the colt does not have a homologue (matching part) on the incomplete Y chromosome. If this region contains the recessive gene for a certain characteristic (such as color blindness or hemophilia in man) and is paired with a Y there is no compensating dominant gene to offset the characteristic; therefore, it is manifest only in the male individual while masked in the female who is the carrier.

Horsemen believe that an ancestor in tail male line exerts more influence than any other ancestor of the same generation. This is true especially in the further generations as the Y chromosome must always be passed from father to son in unbroken line so that a colt receives intact this small portion of his tap root sire's genes, while the chances are only one in about

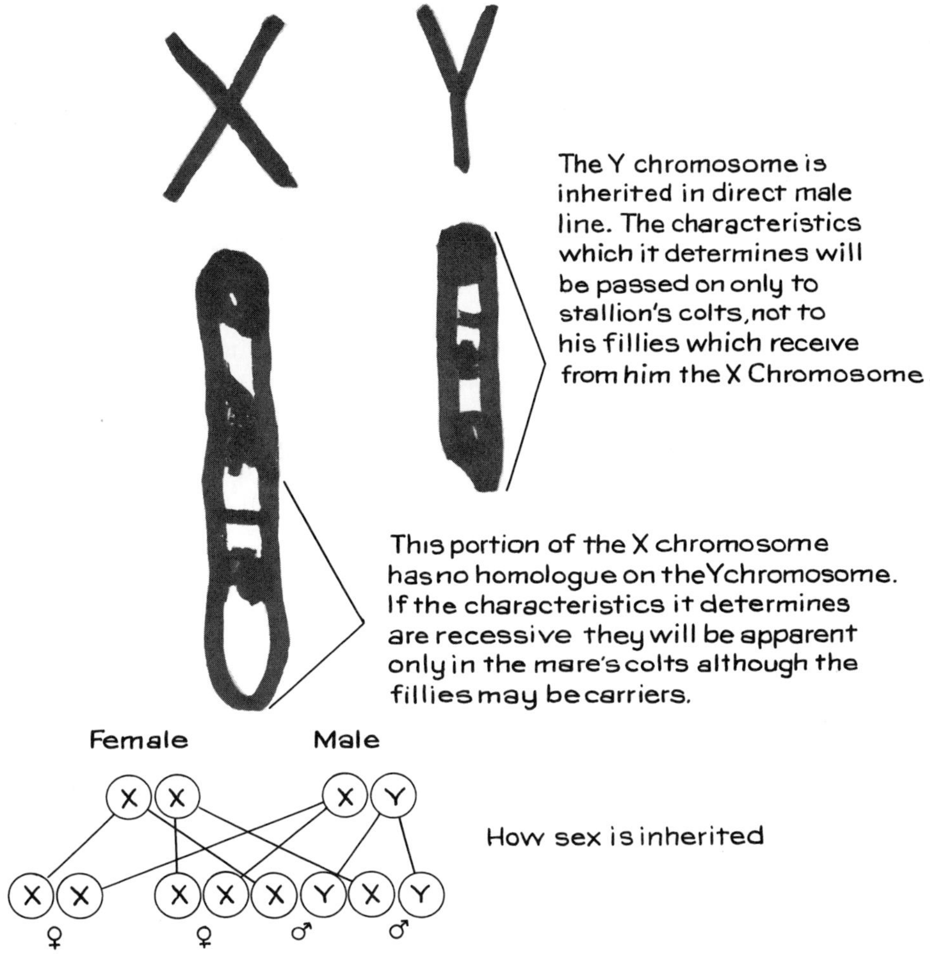

Sexual inheritance: determination of sex, sex-linkage, and sex-linked recessives. The stallion passes on certain traits only to his sons; others, in the X chromosome, pass only to his fillies, which they in turn pass on to half their sons. The Y, however, is passed on to all the sons' colts and to their sons in direct and unbroken inheritance for as long as the particular line endures.

500,000 of his receiving even a single chromosome from another ancestor twenty-four generations removed.[1]

Horsemen also believe that a mare in tail female line exerts more influence than elsewhere on a pedigree; there is no genetic theory to back up this assumption. Although removed by 260 years, Dot Sena, however, bears a striking resemblance to her bottom line tap root ancestor in color, conformation, and marks, and might have posed for the painting of Mr.

[1] In the male line: Night Season is by This Evening, Eight-Thirty, Pilate, Friar Rock, Rock Sand, Sainfoin, Springfield, St. Albans, Stockwell, The Baron, Birdcatcher, Sir Hercules, Whalebone, Waxy, Pot-8-O's, and Eclipse. Eclipse, then, would exert a greater influence than would Woodpecker of the Herod line, also 16 generations removed. Eclipse's influence is even greater in that, of sixteen male ancestors, only two trace back to Herod (through the Tetrarch) and one to Matchem (through the Hastings line), the other foundation sires of the Thoroughbred.

Burton's Natural Barb Mare, the progenitrix of Thoroughbred Family Number 2 (Bruce Lowe).

Some characteristics are the result of incomplete dominance or blending, such as skin color in man and probably palomino color in horses, as well as all the subtle variations in shade of chestnut and bay.

Mutations are changes in the genes themselves brought about by unknown factors either at mitosis or meiosis. They are rare, usually, but not always, recessive, and are often undesirable. The rare white Thoroughbreds foaled in France and Kentucky in 1963 are probably due to mutation as are silver fox and mutation mink, as the name implies.

CHARACTERISTICS DOMINANT AMONG HORSES

Most work has been done with coat color because it is the most obvious feature.

In general, gray is dominant over all the other colors. Therefore gray horses must have at least one gray parent. In Thoroughbreds, all grays trace back in unbroken, though not direct, line to the Alcock Arabian, principally through Roi Herode and his son, The Tetrarch, foaled in 1911 out of the chestnut mare Vahren. In Thoroughbreds, gray and impermanent roan seem to be the same trait genetically with roans producing grays and grays producing roans. Horses originally registered as roan with advancing age are indistinguishable from those described as gray as the white hairs exclude the colored ones.[2]

In theory homozygous grays should produce nothing but gray offspring The Lippizaners of the Spanish Riding School of Vienna are a good example of horses which have been bred for generations for gray color; nevertheless an occasional brown is foaled.

Rare dominant white is epistatic to all colors; it is thought to be lethal when homozygous (Castle).

Bay is dominant over chestnut. It is inherited as a patterning factor.

There are probably two types of black, a recessive homozygous black

[2]The roan gelding Whatitoldyou, stakes-placed winner of 12 races and $89,217, traces back in a chain of gray Thoroughbreds to the Alcock Arabian. He is by the imported horse *Flushing II, by the winner of the English Derby, *Mahmoud, out of the gray mare Mah Mahal, out of Mumtaz Mahal, by the Tetrarch, by Roi Herode, by Le Samaritain, by Le Sancy, out of Gem of Gems, by Strathconau, out of Souvenir, by Chanticleer, out of Whim, by Drone, by Master Robert, out of Spinster, out of Sir Peter Mare, out of Bab, by Bordeau, out of Cygnet mare by Cygnet, out of Blossom, by Crab, by the Alcock Arabian. For 260 years through a maze of 33,554,430 possible ancestors the determinant for gray color has been carried down to the present.

Rare white Thoroughbreds. The filly (left) and colt, owned by Herman K. Goodpaster of Lexington, Ky., were foaled in 1963. Both are by the chestnut Ky. Colonel. The filly's dam was Filly O' Mine, a dark bay by a brown and out of a chestnut. The colt is out of Why Wonder, a chestnut. The odd color is probably due to a dominant mutant gene. After a thorough investigation, the Jockey Club has accepted both for registration. (Photo by J. C. "Skeets" Meadors)

Gray gelding, Whatitoldyou, descends through *Mahmoud and The Tetrarch from the Alcock Arabian, ancestor of all gray Thoroughbreds.

which always breeds true, or yields 100% black offspring when mated with another homozygous black, and a mutant dominant heterozygous black which, when mated with the same, gives one brown or bay foal for every three black ones. Some writers attribute black coat color to the presence of a determinant for black without the patterning (bay) factor. Heterozygous black is dominant over lighter colors. There is some indication that black may work as a concentrator of color, producing brown when crossed with bay in about half the instances or liver chestnut when crossed with chestnut.

Chestnut, or the absence of black points, is recessive to bay patterning. Chestnuts may also have light, flaxen, or pure white manes and tails, inherited as a separate characteristic and superimposed upon any of the many shades of chestnut.

Palomino color is an example of mixed dominance. The coat color is the result of a dominant diluent factor acting on chestnut (absence of bay patterning). Two diluent factors combined make cremello, with pink skin, white mane and tail, cream colored coat, and blue eyes—a semi-albino. The diluent factor paired with bay (patterning) produces buckskin. A mating of palomino to palomino yields, therefore, on the average, one chestnut and one cremello foal for every two golden ones. Palomino to chestnut gives one chestnut for each palomino. Only the combination of cremello, which is homozygous for the diluent factor, and chestnut produces 100% palomino foals. This fact is ironic, as horses with cremello background are ineligible for registration as Palomino horses! It is obvious that palominos, with their heterozygous color-determinant pairings, can never be expected to breed true.

Pinto patterning is inherited separately from coat color and may occur superimposed upon any color. The tobiano type (colored spots on white ground) most common in the United States is dominant over solid color, while the rarer overo type (white spots on colored ground) is thought to be recessive.

If we accept the above principles, including the premise that a diluent factor if present acts as a partial dominant, what accounting is there for the color of Duke of Blackacre, a palomino tobiano out of a dark bay or brown tobiano mare, Dolly D., and by the registered gray Arabian horse Naja? Or for the palomino colt by the chestnut Morgan Devan Marsh Hawk which Duke's full sister, Blackacre's Deborah, a bay tobiano, produced in May of 1964? Is it possible that the diluent factor can also be recessive? Or are some palominos really very light chestnut horses with flaxen manes and tails?

Marks are inherited, according to Castle, as a recessive or a weak domi-

Goldy, a chestnut Saddlebred mare, with her two-week-old palomino colt. The golden coat color is not fully developed until maturity. (Photo by Robert A. Boyajian)

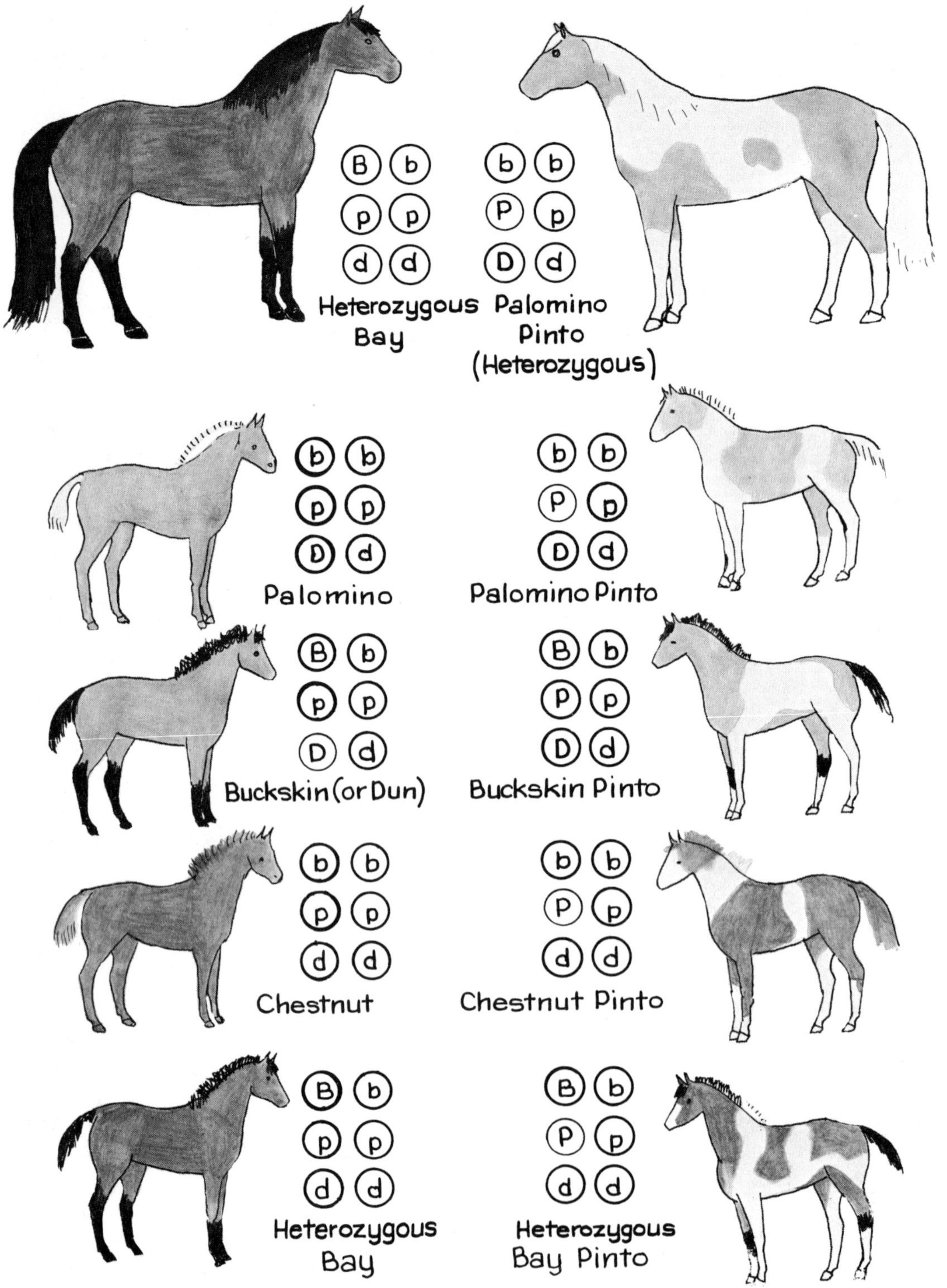

Possible coat color and patterning in foals of heterozygous parents. For each determinant added, the number of possible foals must be doubled. If X and Y chromosomes were included, sixteen rather than eight colts and fillies would have to be shown. If the pinto mare had a glass (blue) eye and the horse carried a recessive for lack of pigment in the iris, the number of possible combinations for these determinants would rise to 32. Considering that there are 33 pairs of chromosomes in the fertilized ovum, the number of possible combinations from any given mating is staggering.

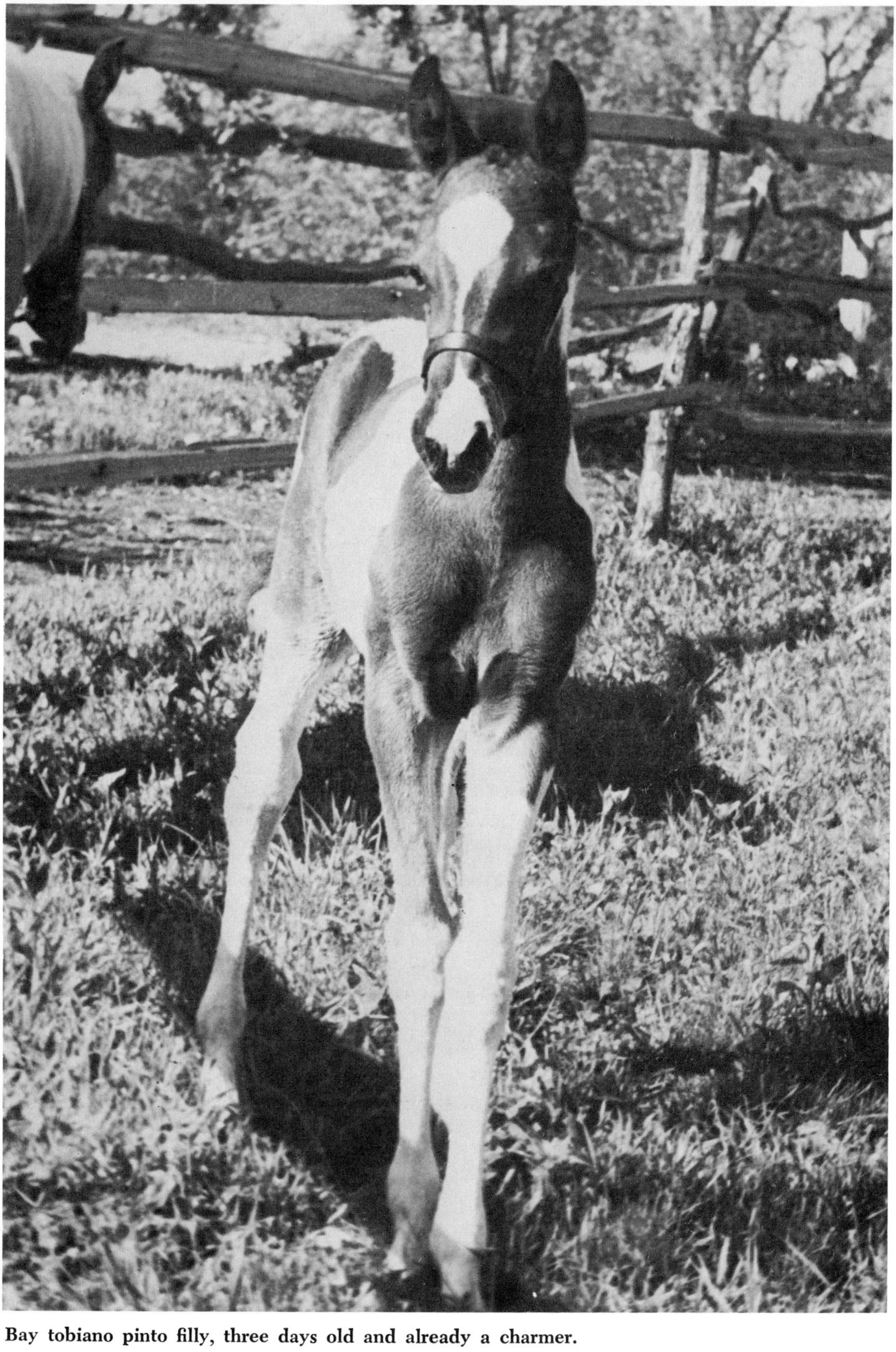

Bay tobiano pinto filly, three days old and already a charmer.

Palomino colt, one day old, by a chestnut horse out of bay pinto mare.

nant. They are quite persistent in many instances. Ethel M. passes on her stockings and wide blaze, the brother and sister Sparko and Esterette being almost identical replicas of their dam. Lady Wentworth describes a case of an Arabian mare with a narrow zig-zag blaze resembling "a flash of forked lightening or a corkscrew" which she passed on to all her foals and which was subsequently passed down through seven generations. A small white spot on the hind quarters of the undefeated Eclipse (1764) was said to have marked all of his many descendents down through the fifth and sixth generations.

Certain general traits seem to be inherited as a group. There is a distinct correlation between eye color and coat color. Pintos often have blue eyes as do bald-faced horses and those with very pale coats. Tetratema, often called the Wonder Horse, a son of the Tetrarch, passed on his spectacular speed only to his gray descendents; his bay and chestnut sons and daughters were not outstanding, with the exception of the unbeatable mare Tiffin.

Horsemen believe that a small, flat-set eye (pig eye) goes with a bad disposition and stupidity. Lop ears, though unsightly, are thought to indicate gameness, competitiveness, and willingness to run.

Ethel M passes on her wide blaze and stockings to her produce.

Esterette, a chestnut filly, has inherited her dam's wide blaze and three stockings.

Most traits, especially those with any economic significance, are inherited in a far more complex manner than has been indicated here. The eye color of the fruit fly, drosophila, is determined by the action of sixty-four genes. Such abstract and compound qualities as racing speed or show ring presence are due to the fortuitous combination and interaction of millions of genes, nurtured and modified by environmental conditions.

Sparko, a chestnut horse, has his dam's blaze and two stockings. (Photo by Hap Farris)

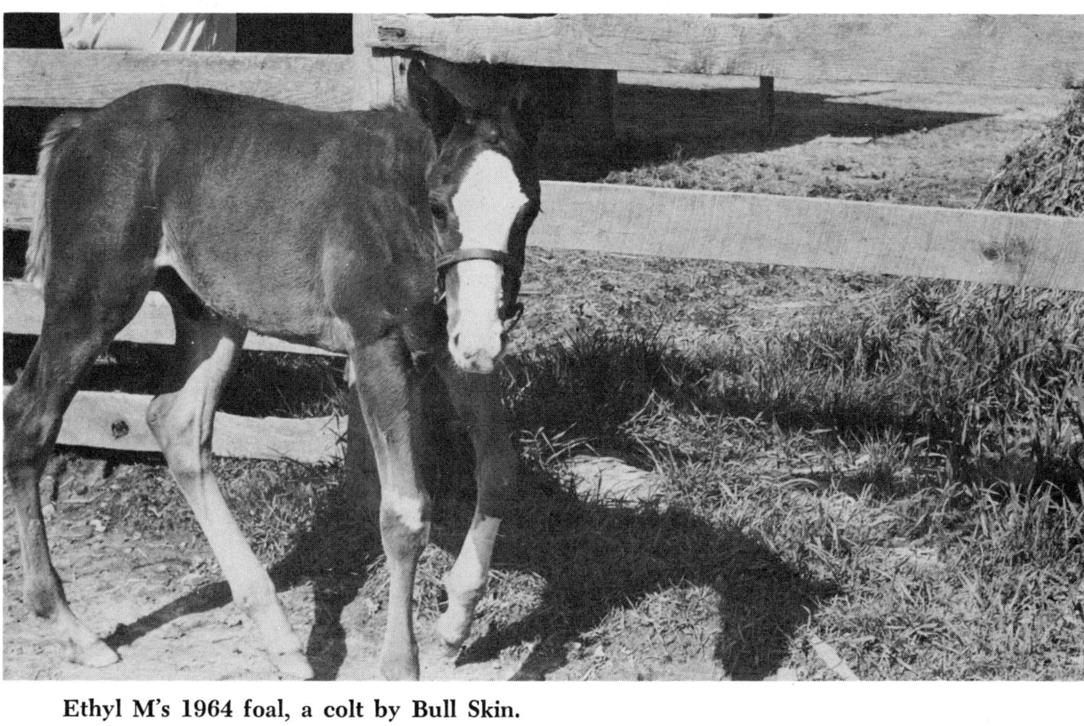

Ethyl M's 1964 foal, a colt by Bull Skin.

14
Registration

IT GREATLY INCREASES THE FOAL'S VALUE TO REGISTER HIM. FAILURE TO DO SO promptly may mean a punitive fee or, if application to register is delayed for too long, the foal may not be accepted for registration at all.

There are as many different rules, regulations, by-laws, and requirements as there are organizations that record horse pedigrees. Anyone who plans to raise horses should familiarize himself with the standards of the breed in which he is interested and should actively support the society dedicated to promoting that breed. In many instances membership in a breed society entitles him to register and transfer stock at a reduced fee. Most registries publish in booklet or pamphlet form their principal regulations, distributed free to those who request it. As rules are continually being revised and amended, the breeder should check directly with the breed secretary rather than consulting material in various other publications which may very well be obsolete by the time he sees it.

The registries distribute promotional literature which includes an account of the breed's development and history, judging standards, present day activities, and photographs of outstanding individuals. An exception is the Jockey Club, which tersely advises those who inquire to consult their public libraries. Usually there is a magazine (or several in the case of popular breeds) devoted entirely to a particular breed. Sometime these magazines are independent; often they are published by the breed society.

Requirements for eligibility to register range from stringent to lax. An animal may be accepted if of the approved color, markings, or type or, in a closed book such as the American Stud Book, which records Thoroughbreds, or the Morgan Horse Register, both parents must be recorded in the register under consideration. The Morgan Horse Register has been closed since 1947. Prior to that time only one registered parent was necessary, the mixed parentage being designated by an X preceding the registration number. The American Stud Book is far stricter. Registration has been denied to an imported Argentine horse because there was an untraced mare in the pedigree seventeen generations back!

Many registries will accept upon inspection an animal which is half-bred. Usually these animals must pass from temporary to permanent status either by inspection at a later date or when the animal has sired or produced the requisite number of approved offspring. Standardbreds may be accepted either on pedigree or performance; the ability to trot or pace a mile in standard time, or to produce standard progeny being considered as important as purebred heritage. Other registries, such as the American Saddle Horse Breeder's Association, will accept half-bred geldings but do not grant the same license to potential breeding stock.

Most registries impose a time limit for eligibility. The deadline for Thoroughbred applications is September 30 of the year in which the foal is dropped. Only before that date will the $30 fee be acceptable. The balance of the first year the fee is $60. In the yearling year, the fee rises to $150 and to $300 the following year, after which the animal may no longer be registered. Morgans may be registered for $20 before December 31 of the year foaled or for $25 thereafter, no longer being eligible for registration after December 31 of the two-year-old year. Members of the Morgan Horse Club enjoy a $5.00 reduction in fees.

Some registries discriminate against an animal because of eye color (Morgan, Palomino) or because of pattern or marks (Morgan, Quarter Horse). In most instances the foal must be registered by its breeder. The breeder is defined either as the owner of the mare at the time the foal is dropped or the owner of the mare at the time she was covered. Other registries allow the owner of the foal to apply for registration but stipulate that the owner of the dam and owner of the stallion both sign the application.

The Jockey Club requires all stallion owners to submit before September 1 a record of all mares bred within the calendar year, the dates on which they were covered, the name of the mare's owner, lip tattoo, etc. A separate service certificate executed by the stallion owner must be attached to all applications. When a mare enters the stud she must be identified to the satisfaction of the Jockey Club and a yearly report made of her thereafter. If she fails to produce a live foal, note of the fact must be made prior to October 31. These checks and balances guard against any possible error or fraud. Usually, but not always. In 1960, a half Percheron filly, Monty's

Babe, was erroneously registered as a Thoroughbred, was sold as a weanling, and underwent extensive race training at Keenland before her patent ineptitude for racing led to a thorough investigation of her background. Registration was revoked.

All registries require a detailed description of the animal's color and marks, which are outlined on a diagram. Several registries require that snapshots of the foal accompany the application.

The color of a foal when it is dropped is often quite different from that of the mature animal. Gray foals are born black, chestnut, or bay, the white hairs not appearing until the first change of coat and becoming more numerous with each moult. Palomino foals are cream colored at birth, the body color barely distinguishable from the pure white mane and tail. Although the skin is pink it darkens to black by the time the foal is a week old. The state of semi-albinism is similar in Siamese cats, which are pure white at birth, developing the seal points as pre-adolescents and the mocha body color as adults. The rich golden palomino coat color is not fully developed until the horse is mature at about five years. Chestnuts are foaled much lighter than they will eventually be. The horn of their hooves is white, pigmented horn being extruded only after birth. As the leg hair shades down almost to white toward the bottom, the presence of coronets, pasterns, or stockings is difficult to detect at an early age. Bays are also considerably lighter at foaling than at maturity. Manes, tails, and hooves (except on legs with marks) are black. Although the black points on legs do not appear until the first change of coat it is possible to predict how high they will extend by the duller brown shade of the leg hair. Browns, except for being less reddish are indistinguishable from bay at birth; the future coat color is suggested in the tone of the body hair, which more closely resembles the leg hair. The dark coat appears with the first shedding and becomes progressively darker until it is actually black except in muzzle and flanks. Brown is one of the most difficult coat colors to identify correctly in the immature individual. *The Racing Calendar* each month carries a list of corrected foal certificates, the greatest number of corrections being in regard to dark bays erroneously identified as "brown" or vice versa. Finally, in 1963, to avoid further confusion the classifications "dark bay" and "brown" were combined in a happy catchall to include all marginal colorations between bay and black.

Blacks are foaled black all over (except for marks) and stay that way, although they may acquire a few gray hairs with advancing age, as do horses of other colors. When sunburned, their coats take on a reddish tinge.

The great majority of white horses were foaled another color, being merely grays of an advanced age. True white horses, which are rare, are

A horse of a different color. Backacre Mandamus, a two-month-old Morgan colt undergoing his first change of coat.

white when foaled. There may be absence of pigment in the skin, in which case it looks pink, and in the iris, resulting in a blue or "glass" eye. A strain of white horses called Albinos (although they are not true albinos) are increasing in number due to the efforts of the late Caleb R. Thompson of Naper, Nebraska who established a registry, the American Albino Horse Club, to promote and encourage the breeding of white horses.

Appaloosa markings may be present at foaling or develop later. The skin is mottled, the hooves are striped, and the horse shows a great deal of sclera in the eye.

Coat color varies with the season of the year. The contrast between winter and summer coat is the most marked in bays and chestnuts. Tippy Dee is dark bronze in the spring, red in the winter. Goldy is copper-colored in the spring, fading to gold by August, and is dark red in the winter. Palominos are much lighter in winter, being cream rather than gold. Nicholls W., a bay, has an almost brown winter coat; his summer hair is thickly roaned, especially around the head, loins, and hocks, giving him a gun-metal appearance. The sprinkling of white hairs appeared at three according to his amended foal certificate, has become more profuse each spring, and yet is almost totally absent in the darker winter coat.

Bays, browns, and chestnuts in high condition are dappled in the spring when the heavy hair is first shed. Later their coats become sunburned, the dappling fades and disappears, and the base color becomes several shades lighter. Show horses, like show poodles, are confined partially to prevent fading so that they will be as dappled in September as in May.

It is unfortunate that crossbred and grade horses do not bring better prices than they do, because they are often outstanding animals. For some purposes a crossbred is superior to a purebred but brings only a fraction of the price unless he has already proved his ability in hunt field, or show ring. The buyer can find many real bargains in hunter, polo, pleasure, and stock prospects among horses and ponies of mixed ancestry.

15
Selling Young Stock

IF THE FOAL IS REGISTERED, BY ALL MEANS ADVERTISE HIM FOR SALE IN ONE of the magazines specializing in the breed in which he is recorded. Even if he is a grade animal of a certain type a breed publication brings him before a large, selected group of readers who are interested in his kind of horse. Many of the subscribers cannot afford to pay the price for a purebred animal but are willing and eager to settle for a typey grade horse of high calibre. It would be pointless, however, to advertise any but a registered Thoroughbred in *Blood-Horse* or *Thoroughbred Record*. These magazines, which are devoted purely to racing and to the breeding and management of potential racing stock, are not concerned even with other Thoroughbred-dominated activities such as hunting, polo, jumping, and combined training. Although *The Chronicle of the Horse* is Thoroughbred-oriented it is a good medium for advertising all breeds of pony, hunters, and hunter prospects of any or no breed, as the magazine is widely read by people who have an informed interest in all kinds of horses. We have had replies to ads from as far away from Ohio as Utah, Rhode Island, Minnesota, Georgia, New York, New Jersey, and Florida.

While advertising in a local paper reaches a larger audience than a breed publication, it is not a selective one. It will bring phone calls and visits from people who are no more than casually interested or who are merely out for a Sunday afternoon's drive.

Public auctions of mixed stock usually do not bring good prices.

Selective sales of registered stock bring excellent prices, as they attract large groups of bidders with a strong interest in the breed of animal being offered for sale. The dates, locations, and conditions of such sales may be obtained from the breed secretary or from one of the livestock agencies.

As a rule, dealers prefer not to buy very young stock, as the demand is greater for "made" horses which are already broken. The amount offered for a foal will be considerably below the animal's actual market value, as the dealer must allow for a reasonable percentage of profit for himself as well as for the cost of boarding the horse for several weeks unless he already has a buyer.

A made horse—one which is already a qualified hunter, polo pony, race horse, or show horse—will sell himself. But the cost of raising and training him to such a degree will often add up to more than the sale price unless the horse is an exceptional individual.

Most trainers are willing to act as agent for a fixed fee or for a percentage of the sale price.

Many farriers know what is for sale, who is in the market for what sort of mount, and the general price range. Rare indeed is the blacksmith who has not acted as dealer from time to time in conjunction with his vocation.

Attention to detail in taking snapshots of the horse to mail to prospective buyers can do much toward showing off his good points while minimizing his shortcomings. Horses are posed with the near foreleg slightly in advance of the off foreleg to give the appearance of a nicely sloping shoulder. As the camera tends to make whatever is nearest appear larger than it actually is, photos of Quarter Horses are taken from slightly to the rear of the animal to accentuate the look of highly developed hind quarters. For a balanced appearance aim the camera at the middle of the horse's barrel. Shots from the front make the horse look grotesquely large-headed, falling away to nothing behind. If the horse tends to be low in the back, standing him in front of a board fence highlights his weakness. Horses do not photograph well in their rough winter coats. The attendant holding the horse should take care not to block out part of the animal. The horse's ears should be pricked forward to give him an alert expression; sometimes this can only be achieved by having an assistant rattle a feed can or throw his hat in the air. The background should be pleasant and uncluttered, with no telephone poles growing out of the horses back.

A conscientious breeder will not sell his foals to just anyone but will make reasonably certain that they will receive decent care, not only for humane considerations but because the appearance and performance of a neglected and abused animal reflects adversely not only upon its present owner but upon the person from whom it was acquired, the specific blood line from which it originated, and the breed of horse in general.

A horse is at the mercy of a cruel, negligent, or ignorant owner far more than a dog or a cat. If a cat does not get enough to eat he can catch a

A crossbred foal. Two-month-old filly sired by a Thoroughbred and out of the Saddle Horse mare Goldy.

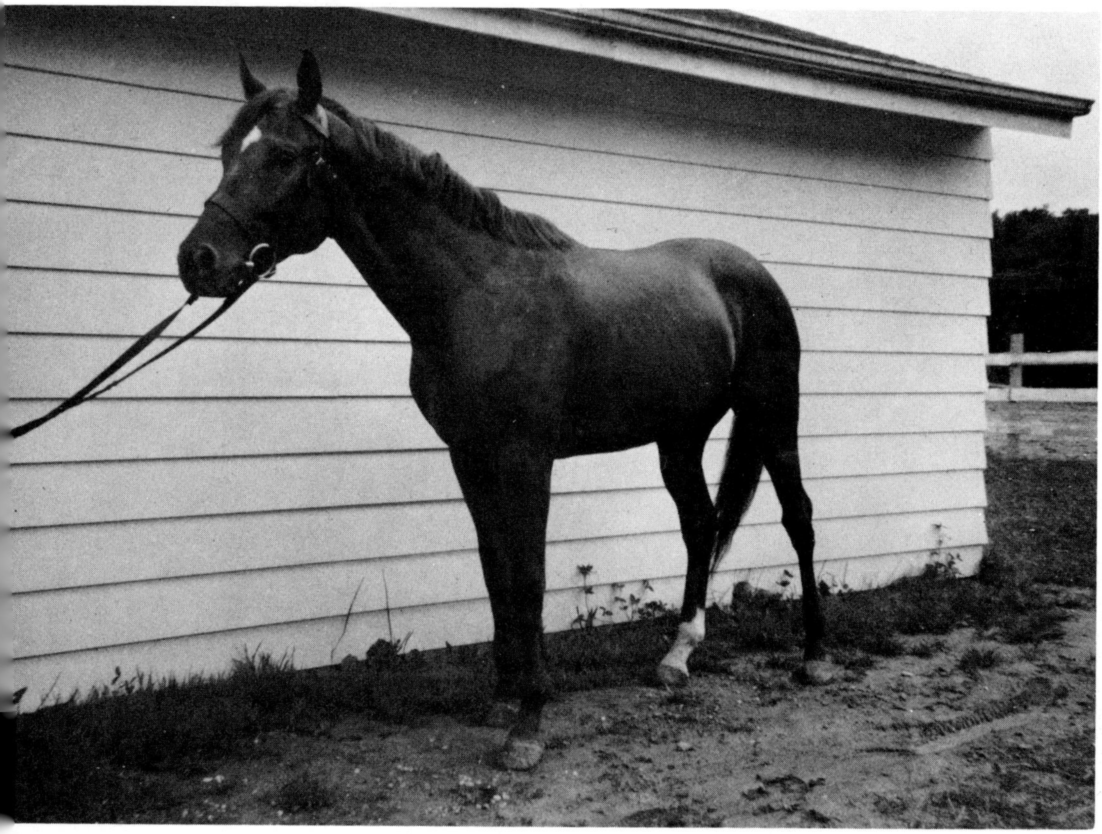

Photos taken from the front make the head appear too large and the hindquarters deficient.

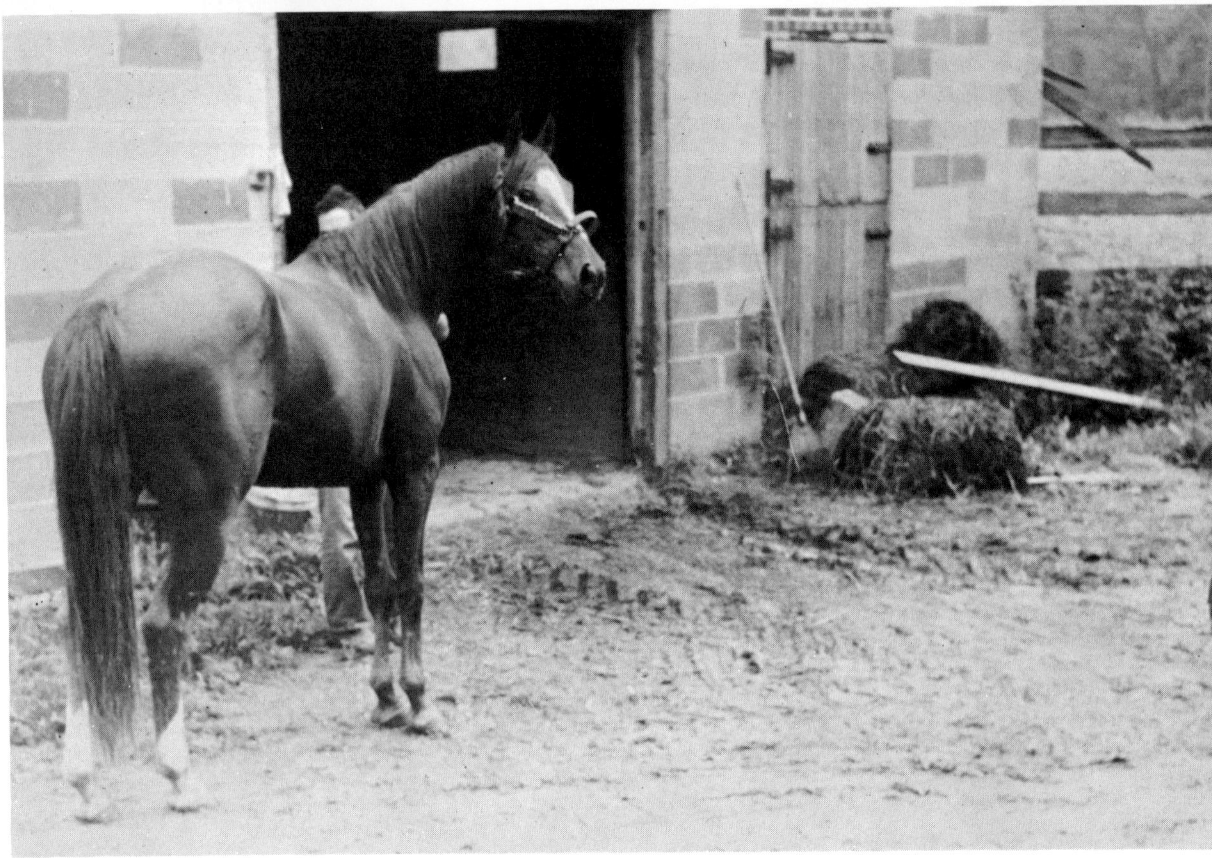

Photos taken from the rear accentuate powerful hindquarters.

An excellent pose. Ears up, near foreleg slightly forward to show off nice shoulder. Parallel lines of background, however, accentuate low back and sloping croup.

mouse; a dog can raid the neighbor's garbage can. If either is too ill-used he is free to look for better pickings. Not so a horse, which is virtually a prisoner and rarely can seek his fortune; only in extreme cases does he have his misfortunes brought to the attentions of a humane officer. It is a cheap form of insurance to give away with each horse sold a copy of the forty-eight page U. S. Government pamphlet *Light Horses* by Dr. M. E. Ensminger, former chairman, Department of Animal Science at Washington State University. (Cat. # A 1.9:2127, 20¢, U. S. Government Printing Office, Division of Public Documents, Washington 25, D. C.)

On the other hand, it is nothing more than a very special form of egotism to suppose that no one can give your stock good care. They may well be better off with someone else.

Appendix A

THE REGISTRIES

Albino

The American Albino Association
Ruth White, Secretary
Box 79
Crabtree, Oregon 97335
1960 registrations: 43
1963 registration certificates issued: 73

Appaloosa

The Appaloosa Horse Club, Inc. Established in 1938.
George B. Hatley, Secretary
Box 640
Moscow, Iowa
1963 registration certificates issued: 11,214

American Appaloosa Association, Inc.
R. Dee Cole, Executive Secretary
Route 4, Box 5
Fort Smith, Arkansas

In 1960 the associations registered a combined total of 4,052 horses.
PUBLICATIONS:
Appaloosa News, Moscow, Idaho
The Appaloosa Breeder's Chronicle, Amarillo, Texas

American Saddle Horse

American Saddle Horse Breeder's Association
C. J. Cronan, Jr., Secretary
929 South 4th Street
Louisville, Kentucky
1960 registration certificates issued: 2,329
1963 " " " : 2,507
PUBLICATIONS:
National Horseman, Louisville, Kentucky. Monthly
Horse World, Louisville. Monthly
Saddle and Bridle, St. Louis, Missouri. Monthly

Arab

Arabian Horse Club Registry of America
Miss Nellie Bayley, Registrar
120 South LaSalle Avenue
Chicago, Illinois
1960 registration certificates issued: 2,104
1963 " " " : 2,892

Half-Arab

Anglo-Arab (Arab-Thoroughbred cross). Included in American Stud Book prior to 1943

The Half-Arabian Stud Book
Ralph E. Goodall, Jr., Secretary
Room 306
224 East Olive Avenue
Burbank, California
1960 Half-Arabian registrations: 2,054
1963 applications received: 5,573
1963 registration certificates issued: 4,803
1960 Anglo-Arab registrations: 67
PUBLICATIONS:
The Arab Horse Journal
Arabian Horse News, Boulder, Colorado. Monthly.
Arabian Horse World, Springfield, New York. Monthly. Profuse use of photographs, many in color.

Cleveland Bay

Cleveland Bay Association of America
Alexander Mackay-Smith, Secretary

White Post, Virginia
Foals registered by British Society.

Connemara Pony

American Connemara Pony Society
Mrs. Bruce Reed, Secretary
Round Robin Farm
East Pepperell, Mass.
1960 registration certificates issued: 103
1963 " " " : 124

Hackney

American Hackney Horse Society
Mrs. J. Macey Willits, Secretary
26 Fair Lawn Avenue
Apt. 1A
Fair Lawn, New Jersey
1960 registration certificates issued: 464
PUBLICATION:
The Hackney Journal, Paullina, Iowa.

Morgan

The Morgan Horse Club, Inc. Registry established 1894.
P. O. Box 2157
Bishop's Corner Branch
West Hartford, Connecticut
1960 registration certificates issued: 1,069
1962 " " " : 1,364. Transfers: 1,408
PUBLICATION:
The Morgan Horse Magazine, Leominster, Mass. Monthly except Jan.

Palomino

Palomino Horse Association. Established in 1936.
Edna Fagan, Secretary
Box 446
Chatsworth, California
1963 registration certificates issued: 125

Palomino Horse Breeders of America. Established in 1941.
Melba Lee Spivey

Mineral Wells, Texas
1963 registration certificates issued: 905
PUBLICATION:
Palomino Horses, Mineral Wells, Texas. Monthly

Pinto

The Pinto Horse Association of America. Established in 1956. Recognized by AHSA in 1963. (The Stud Book was maintained by The Pinto Horse Society from 1941 to 1946.)
Lizann Ayers, Registrar
Somers Road
Ellington, Connecticut
1960 registrations: 225
1963 registration applications received: 985
PUBLICATION:
Pinto Horse News, Ellington, Connecticut. Bi-monthly.

Pony of the Americas

Pony of the Americas Club, Inc.
L. L. Boomhower, Secretary
31 First Street, N. E.
Mason City, Iowa
1960 registrations: 612
1962 " : 693
1963 " : 859
PUBLICATION:
Pony of the Americas Official Magazine, Mason City, Iowa.

Quarter Horse

American Quarter Horse Association. Established in 1940.
Howard K. Linger, Executive Secretary
P. O. Box 271
Amarillo, Texas
1960 registrations: 46,026
PUBLICATIONS:
American Quarter Horse Journal, Amarillo, Texas. Official monthly
Horse Lover, Richmond, California. Monthly
Quarter Horse Bulletin, Fort Smith, Arkansas. Monthly
Quarter Horse Digest, Sturgis, South Dakota. Monthly

Shetland Pony

The American Shetland Pony Club
William Burns, Secretary
P. O. Box 648
Lafayette, Indiana
1960 registrations: 6,897
1963 applications received: 5,346
1963 registration certificates issued: 4,966

Shetland Pony Identification Bureau (records grade ponies).
Daniel Langfeld, Secretary
1108 Jackson Street
Omaha, Nebraska
PUBLICATIONS:
Pony Record, Omaha, Neb. Monthly
American Shetland Pony Journal, 110 West 8th Street, Aledo, Ill. Monthly

Standardbred

United States Trotting Association. Established in 1879.
Don R. Miller, Executive Vice-President
1349 East Broad Street
Columbus 5, Ohio
1960 registrations: 7,124
1961 " : 7,353
1962 " : 7,392
1963 applications received: 7,860
registrations issued: 7,815
PUBLICATIONS:
Hoof Beats, Columbus, Ohio
The Harness Horse, Telegraph Press Building, Harrisburg, Pennsylvania. Monthly
Standardbred Horse Review, Desert Hot Springs, California. Monthly

Tennessee Walking Horse

Tennessee Walking Horse Breeder's Association of America. Established in 1935.
H. Tom Fulton, Executive Secretary
Lewisburg, Tennesseee
1960 registrations: 2,623
1962 " : 3,951
1963 " : 4,250 (approx.)
PUBLICATION:
Voice of the Tennessee Walking Horse. Monthly

Thoroughbred

American Stud Book. Established in U. S. 1829; in Great Britain 1729.
Maintained by the Jockey Club
Anne Holm, Registrar
J. C. Brody, Secretary
300 Park Avenue
New York 22, New York
1960 registrations: 12,625
1961 " : 13,513
1962 applications: 14,808. Foal Certificates issued: Approx. 14,000
1963 applications: 15,536

The Half Thoroughbred Registry
(Formerly maintained by the U. S. Remount Service.)
(Anglo-Arabs recorded in Half Arabian Stud Book.)
Frances C. Hurlbutt
Suite 306C
224 East Olive Street
Burbank, California
1960 registrations: 78

PUBLICATIONS:

Thoroughbred Record and Racing Calendar, Lexington. Weekly.
(*The Racing Calendar,* the official publication of the Jockey Club was incorporated with the *Record* in 1963.)
The Blood-Horse, Lexington. Weekly.
Turf. Monthly.
The Chronicle of the Horse, Berryville, Virginia. Weekly.
(The Thoroughbred in sport. Official publication of the U. S. Equestrian Team, Masters of Fox Hounds Association of America, U. S. Pony Clubs, Inc., U. S. Combined Training Association.)
Racing Form, Triangle Publications, New York and California. Daily.
The Horseman's Journal. Official publication of the Horseman's Benevolent and Protective Association, Danvers, Mass. Monthly.

Thoroughbred Livestock Agencies:
Sales:
Breeder's Sales Division of the Keenland Association
P. O. Box 1690
Lexington, Kentucky
Fasig-Tipton Company, Inc.
250 Park Avenue
New York, New York 10017
Agent: Cromwell Bloodstock Agency
Lexington, Kentucky
Pedigrees, Race Records, etc.:

The Blood-Horse
Thoroughbred Record
Bloodlines, Lexington

Welsh Pony

Welsh Pony Society of America, Inc.
Mrs. Sidney S. Swett, Secretary
P. O. Box 419
Unionville, Chester County, Pennsylvania 19375
1960 registrations: 1,263
1962 " : 1,657
1963 Applications: 1,954
PUBLICATION:
Welsh News, Kennett Square, Pa. Bi-monthly.

In addition to the above more or less generally recognized registries, the October, 1963 all-breed issue of *The Western Horseman* lists the following breed associations and registries:

American Appaloosa Association, Inc.
647 East Lancaster
Fort Worth 12, Texas
National Palomino Breeder's Association
Star Route
Metropolis, Illinois
The American Pinto Horse Association
18832 West Pine Canyon Road
Lake Hughes, California
American Paint Quarter Horse Association
104 National Citizens Bank Building
Abilene, Texas
American Paint Stock Horse Association
Route 4
Gainesville, Texas
International Quarter Horse Jockey Club
P. O. Box 3187
San Angelo, Texas
Model Quarter Horse Association
P. O. Box 1609
Alturas, California
National Quarter Horse Registry
Raywood, Texas
Buckskin Registry Association

P. O. Box 772
Anderson, California
Chickasaw Horse Association, Inc.
Love Valley
Statesville, North Carolina
Galiceno Horse Breeder's Association
711 Peoples Bank Building
Tyler, Texas
Icelandic Pony Club and Registry
#56 Alles Acres
Greeley, Colorado
American Indian Horse Registry, Inc.
P. O. Box 949
Blythe, California
Missouri Fox Trotting Horse Breed Association, Inc.
Ava, Missouri
Morocco Spotted Horse Association of America
Greenfield, Iowa
American Part-Blooded Horse Registry
Suite 107
2073 SW Park Ave.
Portland, Oregon
Spanish Mustang Registry
P. O. Box 148
Finley, Oklahoma
Ysbella Saddle Horse Association
R. R. 2
Williamsport, Indiana

Appendix B

INTERPRETATION OF PEDIGREES

Thoroughbreds

OF ALL HORSES THE MOST CAREFUL BREEDING RECORDS HAVE BEEN MAINTAINED for the Thoroughbred. A certain amount of confusion existed in the early days as horses had the unfortunate tendency to change name with a change in ownership. Sometimes they were known by their owner's name or by the locale in which they were stabled or according to some outstanding physical characteristic which might be variously described depending upon the imagination of the recorder. The Alcock Arabian (1704–23) was also known as Pelham's Grey Arabian, Akaster Turk, The Akastar Arabian, Holderness Turk, Sutton Arabian, Honeywood Arabian and the Brownlow Turk. Mares might simply be known by their sires as Mare by Cygnet, or The Cygnet Mare.

Many systems for classifying lineage have been divised, one of the most widely known being Bruce Lowe's.

Bruce Lowe, an Australian, divided all Thoroughbreds into forty-three families according to the remote tail female ancestor to which they traced. (In fact, there are only eighteen foundation mares, as the others actually trace back to mares previously numbered.) He gave each family a number with Family Number One having the most numerous classic winners among the descendants—which is not surprising, as the greatest number of horses trace to Tregonwell's Natural Arabian Mare. Number Two Family has the second largest number of classic winners, and so on.

> The system is, in fact, a mathematical absurdity when we consider that one single mare on the bottom line of a pedigree of 27 generations . . . represents odds of 27 out of 134,217,728 against her influence and the odds would double themselves with every succeeding generation. She would be entirely negligible unless many times repeated in other lines[1]

[1] Lady Wentworth, *Thoroughbred Racing Stock* (London: George Allen & Unwin, Ltd., 1938) p. 218.

as is Old Bald Peg which swamps most pedigrees.

Although Lowe's system has been subjected to a great deal of ridicule, his numbers are still used to indicate family.

All Thoroughbreds trace in male line to one of three foundation sires Herod,[2] Matchem,[3] and Eclipse,[4] with the last named being dominant. Other males such as the Leeds Arabian, the Sultan Arabian, and the Darcy Turks outweigh the tap root sires in bulk.

Earning index, or the percentage a horse has won in relation to the average expectancy during the years he raced, is the best guide toward selecting potential racing stock. Class tends to breed class. Year after year the descendants of good runners out-earn the sons and daughters of less distinguished performers.

It is not always within the budget of the small breeder, however, to obtain the services of the most eminent stallions. Many of them are syndicated and do not stand to outside mares at any price. The next best solution, always keeping in mind desirable conformation, is to select near relatives of animals which have been notable performers, sires, or producers—preferably including all three. For instance, Quilso (*Princequillo-Maid of Flight by Count Fleet), Kelso's unraced half brother had a full book for 1964 due to breeders acting on the above principal.

This Evening, bay horse, 1952	Eight Thirty	Pilate	Friar Rock	*Rock Sand
				*Fairy Gold
			*Herodias	The Tetrarch
				Honora
		Dinner Time	High Time	Ultimus
				Noonday
			Sea Plane	Man o' War
				*Bathing Girl
	Shimmer	Flares	Gallant Fox	*Sir Gallahad III
				Marguerite
			Flambino	*Wrack
				*Flambette
		Broad Ripple	Stimulus	Ultimus
				Hurakan
			Hocus Pocus	*Sir Gallahad III
				Hazzaza

[2] Great, great grandson of Byerly Turk.
[3] Grandson of Godolphin Arabian.
[4] Great, great grandson of Darby Arabian.

Male line: Eclipse. Family # 6: Old Bald Peg

By Eight Thirty, one of the top race horses of his era, being second only to Horse of the Year Challendon in the handicap division in 1940. Stakes winner of sixteen races and $155,475 in the days before astronomical purses. He has been even more outstanding as a sire, sire of sires, and as a brood mare sire. His more than forty stakes winners include:

Sailor (1952), stakes winner of $321,075. Sire of the stakes winners Bowl of Flowers (best filly of her crop), Ahoy, Crewman, Rocky Link, Sailor Beware.

Bolero (1946), stakes winner of $156,450. Sire of twenty stakes winners including Bolero U, Battle Dance, Jimmer, Tinkalero, Boston Baker, Capelet, Lady Bolero, Tender Morsel, Ben Bolero, etc. Sire of the dams of Dance Lesson, Be Somebody, Petit Rouge, Traffic, etc.

Royal Coinage (1952), stakes winner of five races at two, the only year raced. Sire of the stakes winners Venetian Way (Kentucky Derby), Pierpontella, He's A Pistol, Black Darter.

Eight Thirty is also the sire of the dams of the stakes winners: Evening Glow, Li'l Fellow, Fight Again, Amber Morn ($150,720 sire), Rare Treat

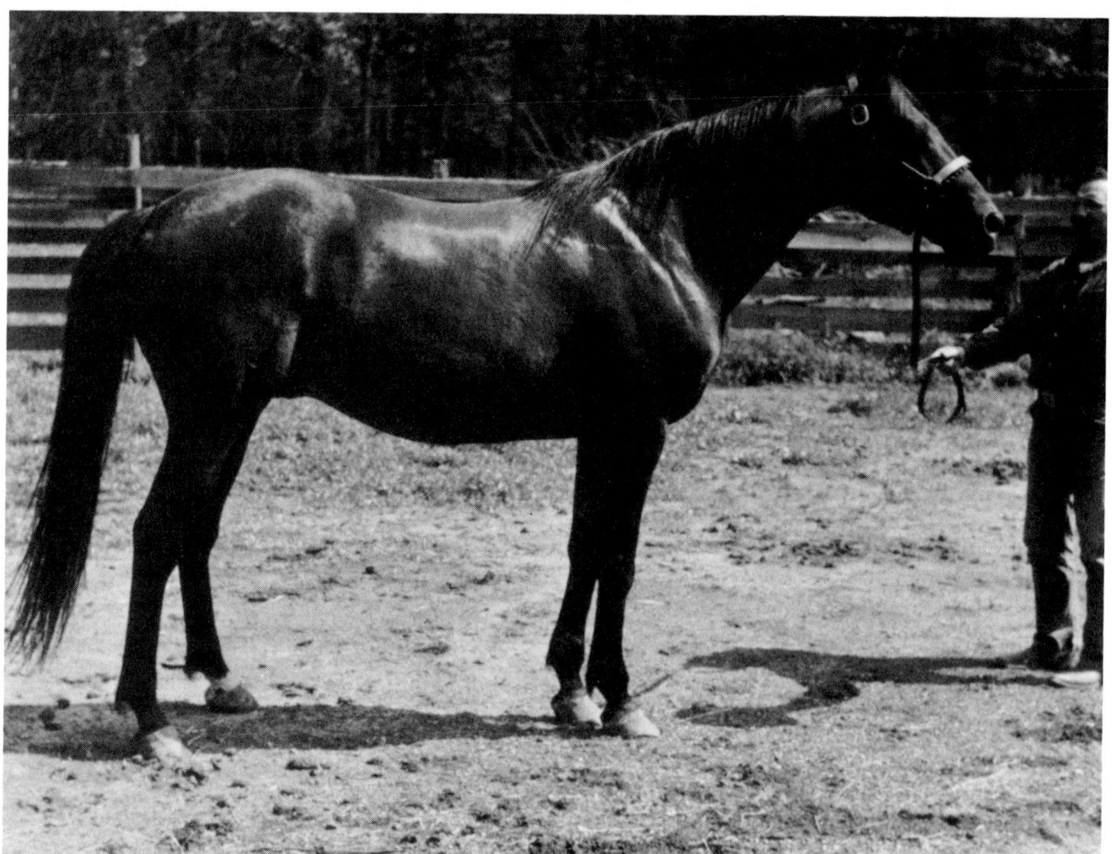

This Evening (by Eight Thirty out of Shimmer by Flares), an excellent example of a mature Thoroughbred. Owned by Roy Vaughn, Long Lane Farm, Newton Falls, Ohio.

($273,227), Jester, Islay Mist, First Balcony, Jaipur ($601,096).

1st Dam

Shimmer (1945) by Flares, brother of Triple Crown winner Omaha. Stakes-placed winner of $23,575. Her foals are:

Orchid (1951) by Black Tarquin. 2 wins at 3.

This Evening (1952) by Eight Thirty. 7 wins 3 to 5. Placed at 7. Sire.

Nantallah (1953) by *Nasrullah. 5 wins 2 to 3. Stakes placed. Sire of Ridan (1959), stakes winner of 13 races, $635,074; Rudoma, stakes winner of $54,910 at 2; Bolinas Boy, stakes winner of 4 races, $26,057; Mount Hope, stakes-placed; Commend.

Prince Shimmer (1954) by *Princequillo, 3 wins 2 to 3. Sire

Royalist (1955) by *Princequillo, 1 win at 2.

Route Twenty (1956) by *Princequillo. Unraced.

Oil Slick (1957) by Oil Capitol. Unplaced at 3.

Shimmy Dancer (1958) by Native Dancer. Stakes winner of 5 races, $37,517.

1959 Barren

*Roman Flare (1960) bay gelding by *Ribot. Placed at 3. 2 wins at 4, 1964.

Minorstone (1961) bay filly by *Ribot. Unplaced at 2.

1962 bay filly by Sailor (which sold at Saratoga in 1963 for $26,000, one of the ten top fillies sold at that yearling sale).

1964 gray filly by Native Dancer.

2nd Dam

Broad Ripple by Stimulus out of Hocus Pocus by *Sir Gallahad III. Stakes winner and dam of stakes winner Rippey, $299,115; Surf Rider, 9 wins; River Flares, 8 wins; Ware On, producer; Royal Ripple; Great Ripple; Sea Ripple; Early Angler; Lois F., producer, dam of stakes winner In Regard.

3rd Dam

Hocus Pocus by *Sir Gallahad III—Hazzaza by Cunard. Unplaced. Dam of

Broad Ripple by Stimulus. Stakes winner of 5 races: Empire Demoiselle Stakes, 2nd in Pimlico Nursery Stakes.

Hoodoo (1941) by Boojum. Stakes-placed winner of 6 races. 3rd Wood Memorial Stakes. Sire.

Sugar Pill. 2 wins at 2. 2nd Betsy Ross Stakes. Producer.

Fake. 12 wins 2 to 4. Stakes placed. Producer.

Wizardess, stakes winner.

Fiddle Faddle

Hokum

Bold Guess

Battle Magic. Producer. Dam of Spy Magic, stakes winner, Oresusuan, stakes placed.

4th Dam

Hazzaza Dam of:
Slipalong, stakes winner
Banter, stakes winner
Zaza, producer
and three other winners

This Evening is an excellent sire for the breeder with a limited budget to consider. He combines a leading male line with a strong family which includes horses that have won or placed in top stakes races. In fourteen seasons Shimmer produced twelve foals or almost double the national average. As a race horse she earned 2.96 times the expectancy. This Evening has a high fertility rate settling most of the mares sent to him, including many maidens and one seventeen-year-old mare which had been barren for five consecutive years. He is prepotent, getting foals of excellent bone, conformation, and size on mares of varied conformation.

Dot Sena, bay mare, 1956	Eternal Road	Eternal Bull	*Bull Dog	*Teddy
				Plucky Liege
			Rose Eternal	Eternal
				Rose of Roses
		Earliana	American Flag	Man o' War
				*Lady Comfey
			Anna Horton	*Wrack
				Royal Message
	Irish Channel	Bold Irishman	*Sir Gallahad III	*Teddy
				Plucky Liege
			Erin	Transmute
				Rosie O'Grady
		Blustery	*Sun Briar	Sundridge
				*Sweet Briar II
			*Windy Way	Junion
				Square Wind

Male line: Eclipse. Family No. 2: Burton's Barb Mare

Eternal Road, stakes winner of twelve races is the sire of the winners Docsena, Dark Road, Dean Sena, Dot Sena, Dion's Pal, Joannie K., Cedar Blen, Eternal Maud, etc.

He is by Eternal Bull, stakes winner and sire of twelve stakes winners including Eternal War, Eternal World, Eternal Bim, Everett Jr., etc. He is the sire of the dams of seven stakes winners including Julmar, Blue Eternal, Cobul, etc.

Eight Thirty, one of the outstanding race horses of his generation, a leading sire, brood mare sire, and sire of sires. He is famous also for his excellent conformation, disposition, and courage. A major influence on the breed. (Photo by J. C. "Skeets" Meadors)

He is by *Bull Dog (1927), stakes winner in France, leading sire in United States, leading brood mare sire three years, sire of more than fifty stakes winners. Sire of the dams of stakes winners Tom Fool, Decathalon, etc.

He is by *Teddy, leading sire in France and the United States. Also sire of *Sir Gallahad III, Teddy's Comet, etc. *Teddy is by *Ajax by the illustrious Flying Fox.

1st Dam

Irish Channel (1947) by Bold Irishman (sire of 1962 stakes winner Plucky Raf). 5 wins 2 to 4. 2nd in Colleen Stakes (one of the big filly races in which females of such class as Cicada and Affectionately also were second in their respective years). Her foals are:

Docsena (1954) by Eternal Road. 10 wins 2 to 7.

Daru's Gal (1955) by Eternal Road. 5 wins at 5 and 6.

Dot Sena (1956) by Eternal Road. 3 wins, 6 times second, third twice, 3 to 6.

Daru-Sena (1960) by Count-A-Bit. Unraced through 1962.

2nd Dam

Blustery (1926) by *Sun Briar. Placed at 2 and 3.

Leeward (1931) by General Lee. 21 wins. Dam of Gallward (1943) winner, Leepap (1944) 4 wins, Dance Wind (1945) 3 wins, Lady Cosward (1947) placed. Dam of Kyward (1956) 6 wins, Prince Pepi (1957) 6 wins, Militant (1948) 16 wins, Buck Private (1949) 9 wins.

Miss Diavolo (1933) 5 wins stakes placed.

Stormscud (1935) 3 wins including Kent Handicap. Sire.

Check Girl (1937) placed at 2 and 3. Dam of Hurry East (1943) 4 wins, Mildred M. (1945) by Stimulus; placed at 3; dam of Top Attraction (1950), 20 wins, Busy Broom (1952), producer, Lew Bro (1954), 7 wins, Last Impulse (1955), 7 wins, Mr. W. R. (1957), 11 wins, Viva's Pride (1958) winner 3 and 4. Marketing (1946), 4 wins. Check Me In (1947), 3 wins. Dam of Dr. A. Pittman (1956) winner, Check Beam (1958) 3 wins.

Choppy Sea (1938) by Hard Tack, 21 wins, stakes placed, $46,810.

Hard Beast (1939) by Hard Tack. 17 wins.

Blow (1941) by *Sir Gallahad III. Placed. Dam of Whiz by (1947) placed at 2; dam of Adaris Queen (1954) by Snark, winner.

Plunge (1948). Unplaced. Dam of Imperial Cadet (1955), 6 wins, $13,572, Mermaiden (1957) 10 wins, stakes placed, Barbara's Jewel (1959), 2 wins, Ho-O-Hum (1960), 2 wins at 2.

Exiled (1950) by Shut Out. Winner.

Stormy (1942) by Hard Tack. 3 wins.

Irish Channel (1947). See above.

3rd Dam

*Windy Way (1920) by Junion. Winner at 2. Dam of:

Blustery (1926) by *Sun Briar. See above.

Sun Blow (1927) by *Sun Briar. 2 wins Stakes. Placed. Dam of Traublow (1933) by *Traumer. Dam of winners Chama Choice, Grand Boy, Fox Blow.

Sun Tempest (1931) by *Sun Briar. 9 wins.

Sun Windy (1940) by Sunador. Unraced.

Dot Sena, although rather on the small side for a Thoroughbred (15.1 h.h.), represents a good family and a strong male line. Every foal but one in three generations is either a winner or a producer and there is more than a smattering of class.

Beverly Blue, bay mare, 1954	Gay Town (bay)	Gay Monarch (roan)	*Royal Minstral (gray)	Tetratema (gray)
				Harpischord
			One Hour	*Snob II
				Daylight Saving
		Dorethea	Nocturnal	Eternal
				Delico
			*Eclipse	*Assagai
				*Hidden Star II
	Medina Blue	King's Blue	Blue Larkspur	Black Servant
				Blossom Time
			Sunny Queen	*Sun Briar
				Cleopatra
		Calanta	*Carlaris	Phalaris
				Carnival
			Donna Santa	Ormondale
				Affable

Male line: Herod. American Family: Mare by *Fearnaught.

Gay Town is the winner of 14 races and $26,770. He is the sire of the winners Gay Marion, Mr. J.T., Gay Bowyer, Mr. Zoar O., etc.

He is by Gay Monarch, stakes winner of 11 races. He is the sire of the stakes winners My Bill, Meagay, Gay Bit, Spring Drift, stakes placed Cadmium, Gypsy Monarch, Merrymarch (3rd Clang Handicap), Gay Guy, and the winners Joe Fisher, Gay Money, Gay Lad, Royal Heels, Gay Orphan, Wit's End, My Pop, Gay Franka, etc. Sire of the dams of stakes winners Montana Rose, Cross Arch and Morpheus.

He is by *Royal Minstral, a leading sire.

He is by Tetratema "the Wonder Horse."

He is by The Tetrarch, one of England's leading sires.

1st Dam

Medina Blue (1946) by King's Blue. Unraced. Her foals are:

Beverly Blue (1954) by Gay Town, unraced.

Night Season (1962) by This Evening.

Blue Hour (1963) by This Evening.

Bay Filly (1964) by High Drag.

Try Blue (1955) by Trymenow. Unraced.

2nd Dam

Calanta (1936) by *Calaris. Winner at 2, 5, 6. Her foals are:
  Medina Blue (1946) by King's Blue. Unraced.
  Bold Cal (1947) by Bold Irishman. Unraced.

3rd Dam

Donna Santa (1922) by Ormondale. 19 wins. Stakes placed. Her foals are:
  Flying Impala (1932) by *Hourless. Unraced. Her foals are: Panziza 1946, winner. Pan Impala (1949), 10 wins.
  High Santa (1933) by High Cloud. 17 wins.
  Ethel Shutta (1934) unplaced. Dam of: Our Chick (1941) dam of Pink Slip (1947), 4 wins. Mesabi (1948), 9 wins.
  Calanta (1936). See above.
  Air Corps (1940), 6 wins.

4th Dam

Affable (1909) by Hastings out of St. Agnes. She is a winner and all of her 8 foals to race are winners, including Lady Madcap (1918) by Dick Finnell, stakes winner of 14 races, 3rd in Kentucky Oaks.

Beverly Blue is a relatively poor risk for producing racing stock. Although good names appear on her pedigree they are several generations removed. She represents a diminishing male line from leading sire to modest stakes winner to winner in the claiming ranks. The gray color ends with Gay Monarch (roan) bearing out Lady Wentworth's assertion that speed is a corollary of gray color in the Roi Herode line, especially those which trace through Tetratema. On the bottom half of the pedigree a strong family seems to have died out almost completely, except for the produce of Beverly Blue; the last other named descendant, Cereal (out of Hasorm Ele, out of Betty Allen, out of Affable) was foaled in 1956 when the dam was twenty-three years old. The number of wins and the age to which the direct descendants of Affable raced (Marmac by Hadagal-Black Cap out of Lady Madcap raced through age 13 winning 26 races) suggests that the family contained a preponderance of geldings.

As she is over 16 hands, has excellent bone, and is an attractive mare, Beverly Blue is worth breeding because her foals would have ready sale value as hunter prospects if they show no talent as runners. Although the odds are not in favor of her producing runners of high calibre stranger things have happened. Carry Back, stakes winner of 21 races and $1,241,165, is out of Joppy, unplaced in 7 starts; the second dam, Miss Fairfax, was unraced. The third dam, War Land, was unplaced. The great Alsab was out of a $90 mare. This is what keeps the gypsies hoping.

It is possible that the owners of Medina Blue and Calanta became discouraged and gave up too soon or that the mares or other foals met with unavoidable accident in no way connected with their racing potential. As a rule, a mare's first one or two foals are not her best. The third through sixth or seventh are usually the most outstanding, with diminishing quality thereafter. There are numerous examples, however, which refute this statement.

Here are a couple of examples of why one should not give up with a mare too soon:

Blackball by Shut Out. Unraced. Her foals are:

Nancy Strieter (1955) by Devil Diver. Unplaced.

Monkey Business (1956) by Tom Fool. Unraced.

Foolish Fancy (1957) by Tom Fool. Unplaced.

(Are you ready to give up and give her away for a pleasure horse? or does she deserve one more chance before you sell her down the river?)

The Axe II (1958) by *Mahmoud. Stakes winner of 13 races and $393,391.

Hester Prynne (1959) by Dedicate. Unplaced.

Malicious (1960) by Helioscope. Stakes winner of 4 races and $101,175;

or,

Level Sands (1949) by *Mahmoud.

Idaho (1956) by Revoked. Unraced. Died at 2.

English Harbor (1957) by War Admiral. Unplaced.

Guadacanal (1958) by Citation. Winner of 5 races; placed in many important stakes.

Kilmoray (1959) by War Admiral. Stakes winner of 9 races, $70,562.

Note: The family notes on Beverly Blue and Dot Sena were prepared by *The Blood-Horse*.

MORGAN BREEDING

The Morgan is America's oldest—and least numerous—breed of light horse. The Morgan contributed heavily to the development of The American Saddle Horse and the Standardbred and to a lesser extent to the Tennessee Walking Horse and the Quarter Horse.

At one time the United States Government took an active interest in Morgans. In various parts of the country stallions were offered at stud gratis to encourage ranchers and farmers to breed a suitable type of cavalry remount horse. The Department of Agriculture maintained a farm near Middlebury, Vermont, from 1907 till 1951 for the breeding of Morgan horses. (Since 1951 the Morgan Horse Farm has beeen managed by The University of Vermont.) The government stud was headed by the black stallion General Gates, descended in direct male line from Black Hawk by Sherman by Justin Morgan. General Gates 666 (1894) was bred by Joseph Battell who not only donated the horse, but gave the land for the farm and compiled and had printed at his own expense the first volume of The Morgan Horse Register. The sons of General Gates, principally Bennington,

Morgan race.

Morgan English Pleasure class.

Morgan Fine Harness class.

which stood at the Government Farm after several years with the Remount Service, Linsley, which went to Kansas, and Red Oak in Texas, carried on his line. Bennington sired the brothers Mansfield, Canfield (sire of the University of Vermont stallion Panfield), Ulysses, and Querido who stood in California.

Government farm breeding saturates most modern Morgan pedigrees.

Unfortunately, Morgan breeders have no clear-cut index to a horse's class such as is available to Thoroughbred and Standardbred fanciers. Evaluation based on show ring performance is highly subjective. It represents opinion and fad rather than fact. Until recent stringent measures were introduced by the Rules Committee of the American Morgan Horse Club,

Morgan Stock class. The rare gray color, less than 1 per cent of total registry, of this horse (Captain Paleface), traces to an imported Arabian.

Manito (by Lippit Mandate out of Vixen), an ideal Morgan stallion in breeding, type, conformation, disposition, and achievement. Here, with Ann Hopkins up, he is competing in the Trail class at the 1963 Morgan Versatility Show, of which he was champion for the second time. Owned by William Hopkins, Green Village, N. J.

Inc., show ring success was too often the result of unscrupulous activities on the part of a few owners and trainers to whom winning was far more important than the horse's welfare, the improvement of the breed, or good sportsmanship.

A more valid measure of a horse's quality than pot collecting is his ability to perform in competition of a less artificial nature, such as hunter trials, combined training events, or competitive trail rides.

In part to establish a standard for improving the Morgan breed, Ethel Gardner of Philadelphia inaugurated in 1960 an all-Morgan Versatility Show. Under AHSA rules 13 classes are presented in which the horses acrue points toward a championship: Stock, English Pleasure, Trail, Fine Harness, Running Race Under Saddle, In-hand, Saddle, Trotting Race in Harness, Western Pleasure, Pleasure Driving, Jumping, Walking Race, and Work Harness. Twice winner of this delightful annual event was Manito, owned by William Hopkins of Green Village, New Jersey.

Manito, chestnut horse foaled June 24, 1949	Lippit Mandate 8211	Mansfield 7255	Bennington 5691	General Gates 666
				Mrs. Culvers by Harrison Chief (ASHR)
			Artemesia 02731	Ethan Allen 3rd.
				Orcutt mare by Bob Morgan 4549
		Lippit Kate Moro 04785	Moro 7467	Welcome 5702
				Polly Rogers 02109
			Croydon Mary	Croydon Prince 5325
				Kate by Radway horse
	Vixen 05128	Virgil 7618	Bennington 5691	General Gates 666
				Mrs. Culvers by Harrison Chief AHSR*
			Quenelda 04268	Troubadour of Willowmoor 6459
				Sunflower Maid 0241
		Althea 04595	Allen H 6527	Ethan Allen 3rd.
				Black Bess
			Agatha 02680	Croyden Prince 5325
				Artemesia 02731

* American Saddle Horse Register.

By Lippit Mandate also sire of: The Third Man, Manitoba, Man o' Day, Man o' Destiny, Chipp-Man, Duke of Marlo, Fortitude, Lippman Hawk, Mr. Showman, The Flying Dutchman, Ann-A-Date, Cherridate, Danny Date, Elane's Date, Flicka Date, Lady Patricia Mar-Lo, Mandate's Peggy Lou, May Date, Nancy Date, Surprise Date, Lady's Man, Hilcrest Lightening, Mandate Encore, Nanana, Mandana, Masterman, Mystery Man, Nobleman, Talisman, Aida Mandate, Carolina, Perfect Date, Polly's Melodate, Penbuck, Commander, Mr. Showman Vona, Penny's Surprise.

Mansfield is a brother to Ulysses, Canfield, and Querido. He is the sire of: Captor, Champlain Banner, Inwood, Joel, Lippit Byfield, Lippit Mandate, Lippit Mormon, Lippit Scofield, Lippit Trilman, Malta, Niles, Osage, Townsend Man, Diana-Mansfield, Ingrid, Jemima, Manitude, Mannequin, Mansphyllis, Manzanita, Mixie, Naida, Noontide, O Leta, Sunny, Andy, Brigadier, Buddy, Charleston, Cherub, Cranford, Damascus, Delmar, Denmark Morgan, Doone, Dusty, Equipoise, Errol, Fieldmarshal, Gay Lad, Gay Mac, Glamorgan, Goldfield, Grenfel, Harvey, Mando, Manscot Glory, Marvel, Monterey, Revere, Ryegate, Scandal, Scovall, Shawnee, Sidney, Silverline, Sonfield, Southard, Splendor, Tiffany, Tipperary, Tunbridge, Ulster, Use Me, Viceroy, Wallace, Wedgewood, Whippit, Whitman, Ashdaisy, Bernice, Winchester, Winfield, Amy Lou, Arissa, Boulder's Sportsman, Boulder's Yankee Boy, Elmira, Evadne, Fawn, Friendly, Gill, Glady, Goldette, Hermina, Lady Benton, Madelin, Manscott Lullaby, Manscot Merrimaid, Manselba, Mansfield's Folly, Ramona, Rhyme, Rhythm, Topaz, Torch Glow, Ukraine, Ursla, Venice, Verbelia, Verona, Vesta, Vienna, Walla Walla, Winsome, Fillmore.

1st Dam

Vixen (1937), bred by Charles A. Stone. An outstanding show horse and winner of the first Morgan Stake at Harrisburg. Destroyed at the age of twenty-six. Because of her extensive show career she produced only four foals:

Manito by Lippit Mandate. Chestnut horse, 1949.

May Date by Lippit Mandate. Bay mare, 1951. Owned by Mr. and Mrs. John Gabbey of Buffalo, New York.

Sherimill Sample by Sherimill Sunrise.

Sunrise Souvenir by Sherimill Sunrise.

Manito is a proven sire. His son A. B. Dillon is also a versatile and distinguished show horse.

He is described by his owner as "a family and neighborhood pet."

Manito represents a solid, high-percentagee pedigree combining Knight (Lippit), Chas. A. Stone, and government farm breeding. In conformation, soundness, disposition, and unlimited willingness to *do*, this little chestnut stallion might well serve as a prototype of the Morgan Horse. He is a qualified hunter and a winner in open as well as breed classes.

Morgan stallion Mansfield. (Courtesy of The Morgan Horse Club, Inc.)

Morgan stallion Bennington. (Courtesy of The Morgan Horse Club, Inc.)

Replevin, chestnut mare foaled August 15, 1959	OCR	Captor	Mansfield	Bennington by General Gates
				Artemesia
			Narissa	Troubadour of Willowmoor
				Ellen
		Roz	Delmont	Ulysses
				Ladelle (AHSR) by Brookwood King
			Ceres	Bennington
				Lady Sealect by Sealect
	Tippy Dee	Hawk Jim	Linsley	General Gates by Denning Allen
				Sunflower Maid by Headlight Morgan
			Lady Spar	Spar Hawk
				May Hudson
		Tippy Tin	Delmont	Ulysses
				Ladelle
			Ceres	Bennington
				Lady Sealect

OCR is the sire of the following registered Morgans: Miller's Chirardo, Lady Esther, Freeman's Rockfire, Reata's Sweet Rhythm, Tas-Tee Black Knight, Tas-Tee Fashion Plate, Tas-Tee Grace, Tas-Tee Indian Summer, Tas-Tee Little Pawnee, Tas-Tee Autumn Haze, Tas-Tee Big Star, Tas-Tee Captain Kid, Tas-Tee Fire Fly, Tas-Tee Jubilee, Tas-Tee Midnight, Tas-Tee Thunderbird, Devan Aire, Devan Comet, Devan Carol, Devan Cavalier, Devan Chris, Devan Dale, Devan De Vere, Devan Fiddlebow, Devan C., Devan Dash, Devan Dimglow, Devan Foam, Devan Hattie, Devan Lady Fireflight, Devan Lady Jane, Devan Mimi, Devan Pamella, Devan Regina, Devan Victoria, Devan Pambroke, Devan Radcliffe, Devan Scot, Devan Wales, Devan Windsor, Devan Gremar Shawn, Devan Harod, Devan Jet, Devan Lennox, Devan Cap, Devan O., Devan Star, Devan Imp, Devan Lady, Devan May Queen, Devan Pat, Darla Dee, Miss Step Along, Replevin, Assumpsit, Cecil's County Fury, etc., etc.

Captor is the sire of: OCR, Fleetwing, Devan Barb, King Captor, Devan Duke, Devan Chief, Devan Comet, Devan Dee, Devan Diamond, Devan King Copper, Devan Prince, Devan Tip, Cappy Rick, Devan B., Devan Circe, Devan Delight, Devan Donna, Devan Duchess, Devan Glo, Devan Gold, Devan Hollycap, Devan Jill, Devan Joy, Devan Queen, Devan Rose, Devan Tess, Devan Val, White Cap, Wingfoot, Caprice, Captor's Princess,

Morgan stallion General Gates. (Courtesy of The Morgan Horse Club, Inc.)

Charm, Starflake, Devan Fuz, Devan Deb, Devan Gleam, Devan Roz, etc.
Mansfield: See above for his get.

1st Dam

Tippy Dee 06510 (1944) by Hawk Jim. Her foals are:

Foxy Prince 11100 (1953) by Captain Fillmore, chestnut gelding.

Cap's Stormy 11309 (1954) by Devan Cap, chestnut gelding. Ohio State Champion Road Hack, Champion Hunter Seat Equitation, Reserve Champion English Pleasure Horse in 1961. Now winning jumping classes and hunted by owner Cameron York of Memphis, Tennessee.

Bar F First Lady 09452 (1955) by L. U. Colonel, dark chestnut mare. Producer. She is the dam of:

Miss Step Along (1961) by OCR, dark chestnut filly. First Weanling Fillies, Reserve Champion Weanling, Ohio Morgan Breeders Futurity, 1961;

Darla Dee (1962) by OCR.

Chico's Flame 11826 (1956) by L. U. Colonel. Chestnut stallion. Won over 70 ribbons in the first year in which he was shown. Sire.

Tas Tee Witch Doctor 12175 (1958) light chestnut gelding by OCR.

Replevin 01613 (1959) by OCR, chestnut mare.

Blackacre Mandamus 15167 (1964) by Devan Marsh Hawk, chestnut colt.

A. B. Dillon, Christine Buzec up. (Photo courtesy of William R. Hopkins)

Tippy Dee and her produce.

Assumpsit, by OCR.

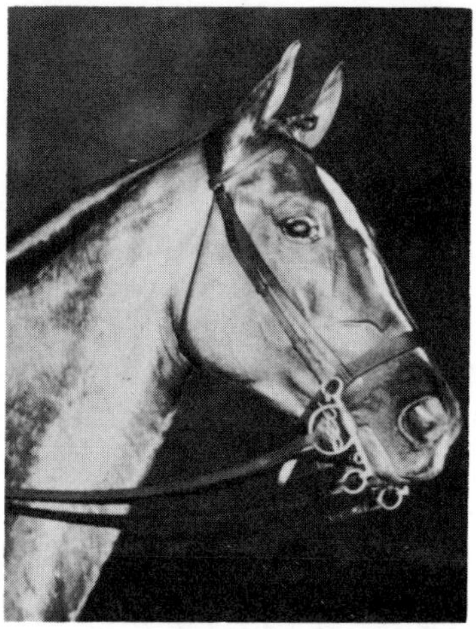

Cap's Stormy, by Devan Cap

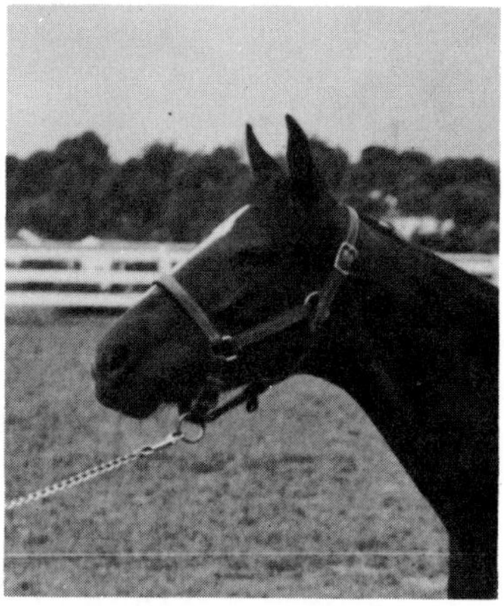

Miss Step Along, by OCR out of Bar F First Lady.

Replevin, by OCR.

Tippy Dee, by Hawk Jim out of Tippy Tin, foaled 1944.

Darla Dee, by OCR out of Bar F. First Lady. (L. U. Colonel - Tippy Dee)

Tas-Tee Witch Doctor, by OCR. Owned by Patricia Wilhelm of Strongsville, Ohio.

Debt, by Devan Marsh Hawk.

Assumpsit 012007 (1962) by OCR, chestnut filly.
Debt 012790 (1963) by Devan Marsh Hawk, chestnut filly.
Detinue 013549 (1964) by Devan Marsh Hawk, chestnut filly.

2nd Dam

Tippy Tin 06298 (1938) by Delmont, bay mare. Her foals are:
Tippy Dee 06510 (1944) by Hawk Jim, chestnut mare.
King Captor 9551 (1946) by Captor, chestnut stallion, Grand Champion Stallion at Michigan State Fair only time shown. Sire.
Devan B. 08290 (1948) by Captor, black mare.
Devan Dimples 11145 (1953) by Captain Fillmore, bay stallion.
Devan C 09385 (1955) by OCR, chestnut mare.
Devan Arden 010742 (1958) by Star Hawk, black mare.
Devan Tabago 01075 (1959) by Star Hawk, chestnut mare.
Devan Tip 10613 (1949) by Captor, chestnut horse. Sire.
Devan Duchess 08390 (1951) by Captor, bay mare.
Devan Fuz 11527 (1955) by Captor, chestnut horse.
Devan Black Comet 11997 (1957) by OCR, black horse.
Devan Anna 010741 (1958) by Silverhawk, bay mare.
Devan Trilby 011352 by Star Hawk.
Devan Duke 11566 (1952) by Captor. Chestnut stallion. Sire.
Devan Del 11565 (1953) by Captain Fillmore, chestnut horse.
Devan Regina 09103 (1954) by OCR. Chestnut mare.

Replevin is a good brood mare prospect because she is a well-developed, attractive individual, full bodied and with ample bone. She combines an outstanding and highly fertile male line with a solid family. There are numerous outstanding individuals on both halves of the pedigree. As she represents very concentrated Government Farm blood lines she may be considered to be homozygous and when mated to an individual of similar background would be expected to produce Morgans of uniformly high type. She is in foal to Devan Marsh Hawk by Devan Duke-Devan Dimglow by OCR which represents even more stringent line breeding to General Gates.

Bearing out the above prediction, which was written in 1963, Replevin produced a very typey chestnut colt on August 11, 1964—Blackacre Mandamus 15167, by Devan Marsh Hawk.

FOLLOW UP

THIS EVENING ENJOYED ONLY MODEST SUCCESS AS A SIRE OF RUNNERS BUT WAS outstanding as a sire of show hunters, getting horses of superb conformation, good size, substance, excellent disposition and way of going. He is the sire of the hunter champions Blue Hour, Quo Warranto (now being shown under the name of Gatsby), Isle of Erin (registered name, Cotton Jeans), and Twelfth Night (registered name Taurus D.).

Through 1971, Beverly Blue produced a live foal every year that she was bred. As predicted, her foals did not do much at the races. Night Season was injured as a yearling and did not race, but is a successful sire with an average earnings index of 3.29 through 1971 which places him in the top 2% of North American sires. His best foal is Burglary, winner of 22 races, 8 of them stakes, and over $125,000. Henri's Pet—formerly Stare Decisis—Beverly Blue's daughter by High Drag, won a few cheap races. Blue Hour and Quo Warranto, never at the race track, earned many honors in the show ring. Beverly's other foals are Blue Eclipse who never won a race but earned her way running second; Forgery, unplaced in five starts, and being reschooled as a 3 day horse; Cerulean Blue being trained as a hunter; Superior Court a two year old and Military Law, a yearling.

Dot Sena is the dam of stakes placed winner of over $14,000, Quick Road, and the winner, Scienter.

There continues to be a strong market for the big, good-looking Thoroughbred either as a race, show or hunter prospect. The steady increase both in racing days and number of races has created a steady demand for Thoroughbreds. The sale as polo ponies, hunters, jumpers, show and pleasure horses insures a valid second market.

Appendix C

PHYSIOLOGY OF OVULATION, FETAL DEVELOPMENT, AND DEVELOPMENTAL ANOMALIES

DURING THE FALL AND EARLY WINTER MONTHS THE OVARIES ARE SMALL, shrunken, and inactive. Heats which occur during this period are usually anovulatory. As spring approaches, and under the stimulus of lengthening hours of daylight, the blood supply to the ovaries increases; the cervix and vagina also become more vascular. As estrogen secretion increases, the winter coat begins to be shed. Ovarian follicles develop and produce lutein. At the beginning of the estrus the follicle reaches a diameter of three to four centimeters. Just prior to ovulation the follicle softens and, after ovulation, the whole ovary becomes flaccid. Ovulation occurs near the end of the heat period. The changes which take place in the ovary may be palpated by rectum to determine the occurrence of ovulation in mares with "silent" estrus.

The ovaries discharge an egg alternately, with the active gland being twice the size of the inactive one. Follicle growth and production of lutein continue through the early months of pregnancy, reaching a peak in the second and third months. When gonadatropic hormone in the mare's blood reaches a high level, the Friedman test for pregnancy becomes possible. Commercially, estrogen is recovered from pregnant mares' urine for human use, 25,000 gallons of urine being necessary to produce one pound of powdered hormone. New follicle formation ceases before the fifth month and hormone production is taken over by the placenta. The transition is a critical period, with imbalance resulting in abortion. From the fifth month the ovaries shrink, becoming small and fibrotic.

At meiosis the ovum does not divide into two equal parts. The chromosomes are equally divided between the mature gamete, which receives the far greater share of cytoplasm, and the tiny polar body which is a by-product, incapable of fertilization. It is the cytoplasm of the mature ovum which nourishes the growing embryo during its initial stages.

After the union of the two haploids the new cell so formed is called a zygote, from the Greek *zygotos* meaning "yoked." After fertilization the zygote divides in the normal manner by mitosis, first into two segments, then four, then eight, sixteen, thirty-two, etc., until there is a mulberry-like cluster, or morula, of nucleated cells. The cells do not continue to multiply at the same rate. Some grow much faster than others. The ovum becomes hollow. The original spherical shape disappears and, as the embryo grows, it elongates, invaginates, and curves upon itself in a canoe-shaped figure. The cells are organized in three layers: the outermost ectoderm from which will come skin, hair, hooves, lens of the eye, nervous and glandular tissue, linings of the nasal cavity, mouth, and anal canal, as well as the fetal membranes, the amnion and chorion; the middle layer of cells is called the mesoderm from which develop the vascular system, muscles, bone, cartilage, connective tissue, lymphatic tissue and the genito-urinary tract; the inner most layer is the endoderm from which come the alimentary canal (except mouth and anus which are ectodermal), the linings of the internal organs, respiratory tract, and the fetal allontis and umbilical cord.

The uterus of the mare is Y-shaped. Fertilization occurs in the fallopian tube, either right or left, but, almost without exception, the ovum crosses to the right horn of the uterus. Even when there are twin fertilizations both develop in the right horn. It is the competition for space and nourishment that causes twin conceptions to abort. Twins are probably more numerous than is generally supposed and are becoming more numerous among Thoroughbreds, composing an estimated 5% of all pregnancies.

Until the fourteenth week of development the embryo receives nourishment from histotrophe, the "uterine milk" secreted by the uterine glands and mucous membrane. The blastocyst floats free in the lumen of the uterus until it grows to fill the cavity. This is in contrast to the human embryo, which becomes imbedded in the endometrium of the uterus at an early stage.

At a primitive stage of development the fetal horse resembles an embryonic pig, or chick, or human. It develops an umbilical cord from the innermost cell layer; water-filled membranees form to cushion the fetus, protecting it from shock and injury; they are the allontis from which the placenta later develops, the chorion which is the outermost layer, and the amnion, the innermost membrane. Villi, minute elongated projections, form on the chorion and imbed themselves in the lining of the uterus when the fetus has grown large enough to fill the lumen. The villi become vascular. The blood of mare and foal is filtered through the placental layer which is reflected over the entire maternal surface. Through the placenta the foal's

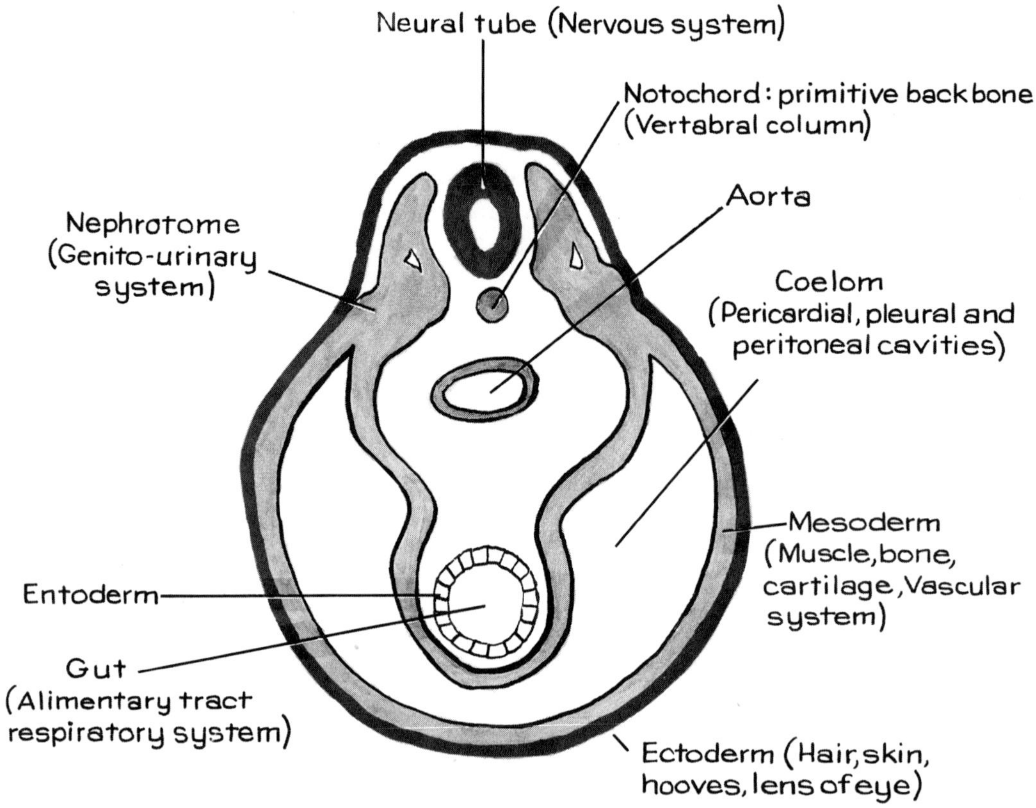

Cross-section of vertibrate embryo at a primitive stage, showing the relation of the germ layers to each other and the origins of various organs and systems.

waste products are eliminated, taken up by the mare's bloodstream and filtered through her kidneys; in exchange the foal receives oxygenated blood and nourishment. Although most harmful substances are barred, certain drugs and viral infections may pass the placental barrier. Fetuses commonly die and are aborted after being invaded by viral rhinopneumonitis or arteritis. Mares which receive anesthesia or sedation during labor produce sleepy foals. Because the placenta is but lightly attached to the uterine endometrium (which is nondeciduate), its separation is easily accomplished without trauma during delivery. Involution is rapid, therefore, requiring only a few days. If labor is prolonged or difficult, however, the foal is likely to die; because of the ease of separation, separation during dystocia is common, the foal being deprived of oxygen from the maternal circulation before it is expelled.

Very early, perhaps as early as the second month, the foal's heart begins beating and the fetal circulation is dependent upon its action from that time; if it is interrupted the fetus dies. Blood is shunted past the lungs via the *ductus arteriosis* which closes at birth, becoming the *ligmentum*

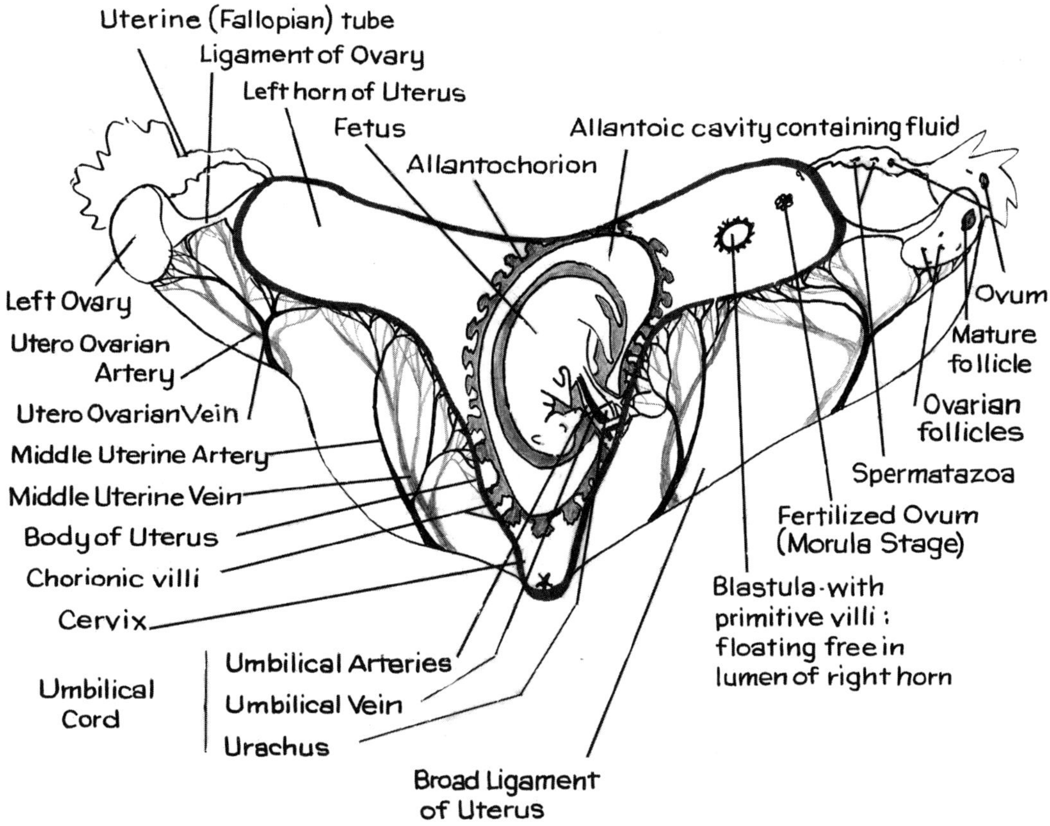

Diagram of uterus, showing ovulation, fertilization, and implantation. The mature follicle ruptures, discharging an ovum which enters the uterine tube, where it is met by spermatazoa. Cell division and differentiation take place rapidly. In actual fact, the horns of the uetrus would be obliterated before the fetus occupied the body of the uterus. Implantation almost invariably takes place in the right horn, even when fertilization occurs in left tube. The uterus is supported by the broad ligaments. The uterine arteries become enlarged and torturous in old mares that have had many foals. Rupture of a uterine vessel is a not uncommon accident in aged multiparae and usually causes rapid death.

arteriosum. A foal with a patent *ductus arteriosis* is cyanotic and weak, usually dying shortly after birth. There are abnormal chest sounds resembling the rumble of machinery. Foals which survive do not grow and develop properly and retain the blueish color of the mucous membranes, due to unoxygenated blood being returned to the arterial flow. Sometimes the anomaly is unsuspected, as in a twenty-year-old horse which was destroyed for use by an anatomy class. Not until the animal was disected was the condition discovered, although he had been thin and had had difficulty in breathing.

The appearance is similar in foals with patent *foramen ovale,* a persistent opening between the auricles of the heart which failed to close normally as the heart developed. The severity of the symptoms is directly proportional to the size of the opening. The foal may have great difficulty in breathing, or be feeble, living only a few hours after birth, although, in less severe cases, the condition is not suspected or discovered until an autopsy is performed, the animal presumably having died from other causes.

Defects in the septum betweeen the ventricles are the most common congenital heart conditions in horses. There are also various malformations of the heart valves which are present at birth or which may be acquired due to scarring and thickening of the valves following generalized infections or by over-use of the immature animal.

Before birth, the fetus secretes urine through the urachus (a narrow tube very much like the ureter), which lies within the umbilical cord. It usually closes at the time the cord is severed, fills with fibrous material and becomes the middle ligament of the bladder. Occasionally the urachus persists. In pervious urachus, urine dribbles from the naval stump, especially when the bladder contracts as the foal urinates. If the condition does

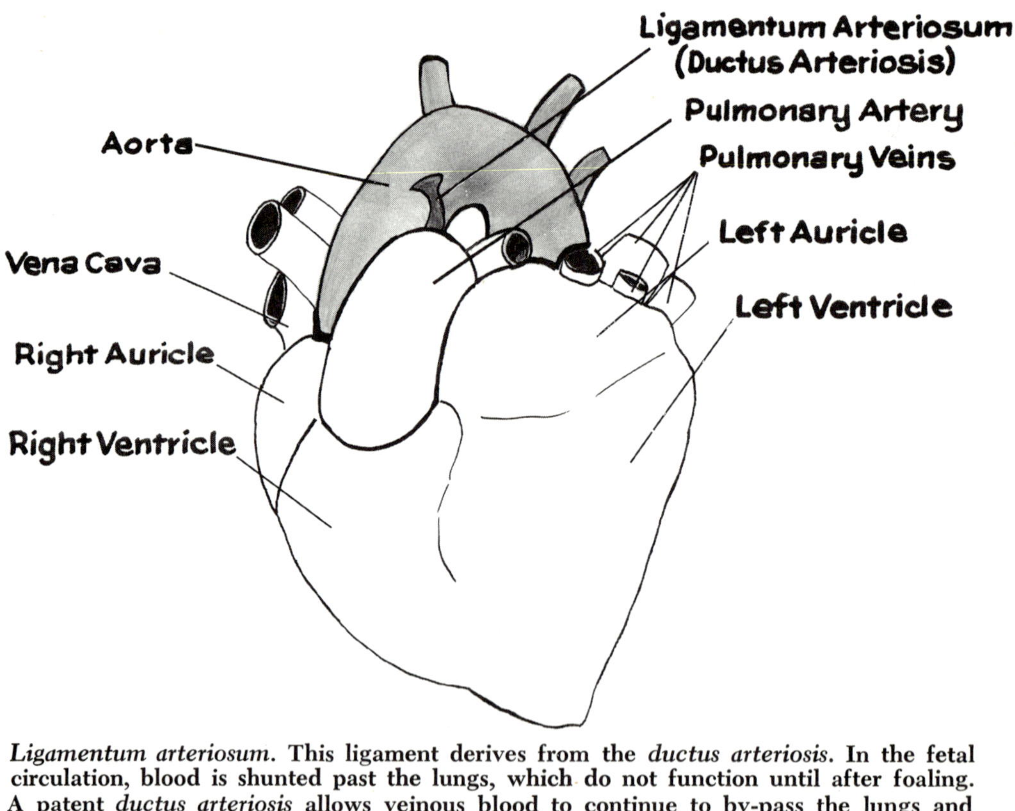

Ligamentum arteriosum. This ligament derives from the _ductus arteriosis._ In the fetal circulation, blood is shunted past the lungs, which do not function until after foaling. A patent _ductus arteriosis_ allows veinous blood to continue to by-pass the lungs and return to the circulation via the main artery, the aorta. A blue tinge to the visible mucous membranes (due to unoxygenated blood) and chest sounds with a machinery-like rumble are characteristic of this defect.

not correct itself in a few days the cord must be cauterized with silver nitrate and cleaned daily until it has healed.

Occasionally there is a defect in the abdominal wall at the site of the umbilicus, through which there is a bulging of gut. Umbilical hernias, if small, are of no significance and soon close by themselves. The Morgan filly Debt was born with an umbilical hernia which had completely disappeared by the time she was a few weeks old. Large hernias are not as likely to close spontaneously. Sometimes the application of a blister promotes closing; others may be clamped while some must be corrected surgically.

In colicky symptoms of male foals, scrotal hernia should be considered as a possible factor. The testicles are formed in the abdomen and pass through the inguinal ring into the scrotum, the age of descent being a highly variable factor. A relaxed inguinal ring may allow a loop of gut to herniate through it causing intense pain. If the hernia strangulates it is an exceedingly grave condition requiring immediate surgical intervention if the colt is to be saved. Although most scrotal hernias correct themselves in time, others must be corrected surgically. Sometimes the testicles are permanently retained in the abdomen or in the inguinal canal rendering the animal sterile, as spermatogenesis does not take place at body temperature but requires the lower temperature of the scrotum. If both testicles are retained the horse is a double cryptorchid; if only one fails to descend, usually the left, he is called a single cryptorchid. A horse which has had the one most easily accessible testicle removed is a monorchid, ridgeling, or rig (although one hears horses with all degrees of cryptorchidism referred to as rigs). Single cryptorchids are fertile but should not be used for breeding as the condition is hereditary.

Some foals are born with a section of gut missing. Several feet of large intestine or colon may be absent, the portions which exist being in two separate pieces each ending in a blind pouch. This condition known as *atresi coli* (from the Greek *atresos*, "not perforated") is usually fatal. Although most of these foals seem normal at birth, are vigorous and able to nurse, they soon develop symptoms of colic which increase in severity in spite of any treatment. Failure to pass feces or fluid after an enema is pathognomonic. Attempts to approximate the ends of the colon are unsuccessful because too much of the intestinal tract is missing.

Hydrocephalus (from Greek *hydros* = water + *cephalos* = brain) is caused by obstruction of any of the small structures which drain the cerebro-spinal fluid. The abnormal and excessive fluid pressure causes a huge head and atrophy of the brain. The distention may be so great as to make delivery difficult or impossible unless a craniotomy is performed. In other cases the foal is essentially normal in appearance but is unable to stand or is listless and uncoordinated. As there is no cure, euthanasia is usually performed on those foals which are not stillborn.

Far less common than foals with the above defects are monsters, which show freakish, distorted development:

Monsters may be divided into—
(1) Monsters with absence of part—absence of head, leg or other organ.
(2) Monsters with some part abnormally small—dwarfed head, leg, trunk, etc.
(3) Monsters through some unnatural division of parts—cleft head, trunk, legs, etc.
(4) Monsters through absence of natural divisions—absence of mouth, nose, eyes, anus, etc.
(5) Monsters through fusion of parts—one central eye, one nasal opening, etc.
(6) Monsters through abnormal position or form of parts—curved spine, face, legs, etc.
(7) Monsters through excess of formation—enormous head, extra digits, etc.
(8) Monsters through imperfect differentiation of sexual organs—hermaphrodites.
(9) Double monsters—double-headed, double-bodied, extra legs, etc.

The cause of monstrosities are varied. Some monstrosities resulting from extra digits, absence of tail [as polydactilism in cats, taillessness in Manx breed of cat], etc., run in families. Others are associated with too-close breeding [q.v.], the powers of symetrical development being interferred with, just as in other cases a sexual incompatibility is developed, near relatives failing to breed with each other. Mere arrest of development of a part may arise from accidental disease of the embryo; hence vital organs or portions of organs, such as the dividing walls of the heart [patent *foramen ovale,* interventricular fault] are omitted. Sometimes an older fetus is enclosed in the body of another, each having begun independently from a separate ovum but the one became embedded in the semi-fluid mass of the other and developed there simultaneously with it but not to so great a size nor so perfectly. In many cases of redundance of parts the extra part has manifestly developed from the same ovum and nutrient center with the normal part to which it remains adherent. In the early embryo, with its great power of development, this factor can operate to far greater purpose than in the adult animal. Its influence is seen in the fact . . . that such redundant parts are nearly always connected with the corresponding portions of the normal fetus. Thus superfluous legs . . . are attached to normal ones, double heads or tails are connected to a common neck or rump, and double bodies are attached to each other by corresponding points, naval to naval, breast to breast, back to back. All this suggests the development of extra parts from the same primary layer of the impregnated and developing ovum. The effect of disturbing conditions in giving such wrong directions to the developmental forces is well shown in the experiments of St. Hilaire and Valentine in breaking up the natural connection in eggs and thereby determining the formation of monstrosities at will. So in the mammal, blows and other injuries that detach the fetal membranes from the wall of the womb or that modify their circulation by inducing inflammation are at times followed by the development of a monster.[1]

[1] James Law, "Diseases of the Genital Organs," *Special Report on Diseases of the Horse* (Washington, D. C.: United States Government Printing Office, 1947) p. 181-2.

Glossary

Actinobacillus. Bacteria causing joint-ill syndrome in young foals. In older foals become localized in soft tissue already invaded by parasites.

Allontis. The chief agent of early fetal circulation, mesodermal in origin; the umbilical cord and bladder develop from it.

Anestrus. Period between heats. Failure to come in heat may be due to pregnancy, season of the year, old age, debility, or surgical removal of the ovaries.

Aneurism. A ballooning out of an artery wall. Sudden death may be due to the rupture of an aneurism of a major vessel.

Anovulatory. Not producing an egg, or mature ovum.

Back cross. The crossing of an unknown phenotype, such as bay coat color, which may be homozygous or heterozygous, with a known homozygous recessive, as chestnut coat color, to analyze genetic content.

Barren. Not in foal. Or not capable of producing a foal.

Blastocyst. A blastula, the fertilized ovum developed into a hollow sphere of cells.

Blastomere. Cells arranged around a central, fluid-filled cavity.

Blastula. A hollow sphere of cells.

Breeding roll. A padded roll used to protect maiden or very small mares from coital injury.

Carpus. The knee. Comparable to the wrist in man.

Capillaries. Very small, hair-like blood vessels which form a network of connecting veins and arteries.

Cecum. The blind gut located at the juncture of the illium and colon.

Cervix. A neck, particularly the neck of the uterus.

Cervitis. Inflamation of the cervix.

Colostrum. The first milk which is laxative and rich in antibodies.

Craniotomy. The reducing in size of the foal's head by cutting or crushing where delivery is otherwise impossible.

Crossbred. The foal of two purebred animals of different breeds as Thoroughbred-Percheron, Anglo-Arab, Arab-Welsh, etc.
Cytoplasm. The protoplasm of a cell other than the nucleus.
Diploid. A somatic cell containing 66 paired chromosomes.
Distal. Away from a central point; peripheral to.
Dystocia. Difficult labor.
Electrolyte. Body salts and sugars.
Embolism. The occlusion of an artery by an embolus.
Embolus, pl. emboli. Foreign matter in the bloodstream; a blood clot, parasite, bacteria, air, etc.
Embryotomy. The cutting up and removing piecemeal of a foal which is hopelessly wedged or is too large to be delivered normally.
Encephalomylitis. Sleeping sickness.
Endemic. Peculiar to a region, occurring more or less constantly in a particular locale.
Endometrium. The mucous membrane lining of the uterus.
Epiphysis. A cap of bone attached by cartilage to another bone, becoming united to the main bone at maturity after which no further growth is possible.
Epistatic. Dominating or suppressing the effect of another gene.
Estrogen. Estrus-producing hormone.
Estrus, heat. Mating period. Silent estrus is the occurrence of ovulation without outward manifestation.
Family. A horse's female relatives and ancestors.
Fibroblasts. Immature, primitive fibrous or connective tissue cells present in early stage of wound repair and scar formation.
Float. To file down irregularities in molar teeth, care being taken not to smooth out the normal grinding surface.
Follicle. A small secretory cavity or sack; in the ovary, containing the ovum.
Gamete. Reproductive cell, either male sperm or female ovum.
Genotype. The assortment of genes in cells.
Gonad. Organ of reproduction, either male testes or female ovary.
Gonadatropic hormone. Hormone which stimulates the gonads.
Grade, unregistered. Usually an animal with one registered parent, or of a distinct breed type, as grade Morgan or grade Saddlebred. A grade horse may be anywhere from a half to 63/64 pureblooded.
Halfbred, halfbreed. Having one registered parent.
Haploid. The mature gamete containing, in horses, 33, half the normal number of chromosomes.
Heat. See Estrus.
Hemapoietic. Blood-making.
Heterozygous. Having mixed genes, not homozygous; hybred. (From Gr. *heter* = "other" + *zygotos* = "yoked".)
Homozygous. Having similar genes. (From Gr. *homo* = "same" + *zygotos* = "yoked".)

Hormone. Excretion of a gland or organ, transported by the bloodstream, and having a specific effect in regulating the activity of cells remote from its origin.

Humerus. Long bone from shoulder to elbow.

Impaction. The state of being lodged and retained in a part, specifically a mass of feces in the intestines.

Invaginate. To infold to form a hollow space; especially, in embryology, the infolding of a part of the blastula to form a gastrula.

Kigme. A kickable character from the world of Al Capp.

Libido. Sexual drive.

Line. A horse's direct male ancestors.

Long yearling. A colt or filly near the end of its yearling year. "Coming two," "rising two."

Lutein. An internal secretion of the ovary.

Luteinizing hormone. Pituitary hormone which stimulates ovaries.

Maiden. Indicates a state of anticipation; *i.e.*, a horse which has not yet won a race, a mare which has not yet been bred, or one which has not previously produced a foal.

Malpresentation. Abnormal position of fetus at foaling.

Mastitis. Inflamation of the udder.

Meconium. Content of the foal's intestines at birth.

Meiosis. Cell division by chromosome reduction; characteristic of gamete just before it matures.

Mesentery. That part of the peritoneum (lining of the abdominal cavity which also surrounds the viscera) which connects the gut with the abdominal wall.

Mesenteric arteries. The arteries which supply the intestines. (Two strongyle larvae were migrating through a mesenteric artery when the vessel branched at right angles. One strongyle turned off but the other continued, as he was afraid to change streams in the middle of a horse.)

Metritis. Infammation of the uterus.

Mitosis. Normal cell division in which the number of chromosomes in each cell remains constant.

Morula. The mulberry-like cluster of cells in the early embryo.

Motility. The ability to move spontaneously.

Multipara, pl. multiparae. An in-foal brood mare which has already produced one or more foals.

Nondeciduate, not cast off. Applied to the endometrium, or lining of the uterus, which is but lightly fused with the fetal membranes.

Nymphomania. Excessive sexual desire of a mare.

Ovum, pl. ova. Egg.

Parturient. Pertaining to foaling.

Pathognomonic. Characteristic or typical of a disease or vice. Said of symptoms which distinguish the condition from other conditions. (From Gr. *patho* = "suffer" + *gnome* = "a judgment.")

Periosteum. A tough, glistening white fibrous membrane which covers bone surfaces, except at the points at which ligaments and tendons attach and on the articular, or joint, surfaces where cartilage is substituted.

Phenotype. The appearance of a horse or other animal.

Post mortum. After death; especially an examination to determine the cause of death. An autopsy. Should be performed in all unexplained deaths.

Postpartum. Following foaling.

Postparturient. After foaling.

Progesterone. The hormone found in corpus luteum.

Return, return privileges. Repeated service or cover, given gratis if mare fails to conceive on the first breeding. May be limited to once or to the current season.

Sclera. White of the eye.

Septum. A partition or dividing wall between two spaces or cavities.

Sibling. Brother or sister.

Spermatazoon, pl. spermatazoa. Sperm; male gamete or reproductive cell.

Spermatogenesis. The production of sperm; sometimes restricted to mean meiosis in the male.

Stifle. Patella. Comparable to the kneecap in man.

Stud. A farm for raising horses. The group within a breed used for propagation; *i.e.*, a mare enters the stud when she is bred, or when she produces a foal. A stallion at public service is said to be at stud. Term is loosely used for any male entire.

Syndrome. A grouping of symptoms, traits, or characteristics which tend to occur together, or which, when they do occur together, are considered to be characteristic of a disease or of a disease process.

Tarsus. The hock. Comparable to the heel in man.

Testosterone. Male hormone.

Thrombosis. Formation of a thrombus.

Thrombus, pl. thrombi. A blood clot due to slowing of the circulation or to alteration of the blood or vessel walls as in parasite infestation. (From Gr. *treiphein* = "curdle" or "congeal.")

Trauma. Wound, injury; shock.

Typey. Typical. Having breed characteristics to a marked degree.

Umbilicus. Navel.

Uterus. Womb; female organ of gestation which contains the fetus during its development and is the principal agent in its expulsion during foaling.

Vaginal speculum. An instrument for dilating the opening so that the interior is more easily visible. Used for observing and taking smears from cervix and vagina.

Vascular. Consisting of, or pertaining to, or provided with, blood vessels.

Vector. An insect, or other arthropod, which carried micro-organisms from an infected to a healthy animal.

Verminous. Pertaining to, or caused by, parasites.
Version. Manipulation of foal to change its presentation during difficult delivery.
Zygote. The fertilized ovum. (From the Gr. *zygotos* = "yoked.")

Bibliography

Aga Kahn, "Foreword," *Thoroughbred Racing Stock* by Lady. Wentworth (London: George Allen & Unwin, Ltd., 1938).

American Morgan Horse Register (New York: The Morgan Horse Club, Inc.) Vol. IV–VIII.

Andrist, Freiderich, *Mares, Foals and Foaling*, tr. Anthony Dent (London: J. A. Allen & Co., Ltd., 1959).

Arey, Leslie Brainerd, *Developmental Anatomy* (Philadelphia: W. B. Saunders Company, 1954).

Barron, Charles N., D.V.M., Ph.D., "Gestation Periods" (letter), *The Thoroughbred Record* (Lexington: 1964).

Bone, J. F., D.V.M., M.S., *et al.*, *Equine Medicine and Surgery* (Wheaton, Illinois: American Veterinary Publications, 1963).

Bowen, Ed, "Face Flies May Become Problem for Horsemen," *The Blood-Horse* (Lexington, 1963) p. 1531.

Bower, Alex (Director), *Lectures: Stud Manager's Course*, The Stud Manager's Course (Lexington: 1963).

Britton, J. W., D.V.M., "Wound Treatment," *Equine Medicine and Surgery* (Wheaton, Illinois: American Veterinary Publications, 1963).

Burch, Preston M., *Training Thoroughbred Horses* (Lexington: The Blood-Horse, 1953).

Castle, W. E., "The A B C of Color Inheritance in Horses," *Genetics* (1948) vol. 33, no. 1, pp. 22–35.

Catalogue: Keeneland Fall Sale (Lexington: Breeder's Sales Company Division of the Keenland Association, Inc., 1963).

Davidson, A. H., D.V.M., "Unsoundness," *The Blood-Horse* (Lexington, 1961) Vol. xxxii, no. 15, p. 902.

Doll, E. R., D.V.M., "A Planned Infection Program for Controlling Abortion," *The Blood-Horse* (Lexington, 1962) p. 1214.

Drudge, J. H., D.V.M., ScD., "New Development in Parasite Control," *The Blood-Horse* (Lexington, 1964) p. 132.

Earle, I. P., "Grassland Crops as Feed for Horses," *Grass* (Washington: U. S. Department of Agriculture, U. S. Government Printing Office, 1948).

Ensminger, M. E., Ph.D., *Horses and Horsemanship* (Danville, Illinois: The Interstate Printers and Publishers, Inc., 1963).

Estes, J. A. and Joe H. Palmer, *An Introduction to Thoroughbred Horses* (Lexington: The Blood-Horse, 1949).

Estes, J. A., "He Cast a Long Shadow," *The Blood-Horse* (Lexington: 1962) p. 758. [Obituary of *Mahmoud.]

———, "Three White Thoroughbreds," *The Thoroughbred Record* (Lexington, 1963) p. 1565.

———, "The St. Simon Heritage," *The Thoroughbred Record* (Lexington: 1964) p. 1402 ff.

Harris, Harold J., M.D., F.A.C.P., *Brucellosis* (New York: Paul B. Hoeber, Inc., 1950).

Heintz, C. D., D.V.M., "Epiphyseal Stapling," *The Thoroughbred Record* (Lexington: 1964) p. 1415 ff.

Hollingsworth, Kent, ed., "Stakes Winners of 1963," *The Blood-Horse* (Lexington, 1964).

———, "2,267 Yearlings Sell for $13,126,925," *The Blood-Horse* (Lexington: 1963) p. 1774.

"A 1,600-Pound 2-Year-Old Filly—No Sale," *The Blood-Horse* (Lexington: 1963) p. 1471.

Humphries, John O., "Trend Away from *Caveat Emptor* Observed in Laws of Horse Sales," *Blood-Horse* (Lexington: 1963) p. 1777.

J., M.P., "Mare, 31, Has Foal," *The Chronicle of the Horse* (Berryville, Virginia: 1964).

Kays, D. J., *The Horse* (New York: A. S. Barnes & Company, Inc., 1953).

Lieux, Pierre, D.V.M., "Infertility in the Stallion," *Equine Medicine and Surgery* (Wheaton, Illinois: American Veterinary Publications, 1963) p. 633.

Mackay-Smith, Mathew, D.V.M., "The Care of Wounds," *The Blood-Horse* (Lexington: 1963) p. 1212.

McClung, Marion, "Florida 100-Mile Trail Ride," *Arabian Horse World* (Springville, New York: 1962) Vol. II, no. 10.

Metcalf, John W., D.V.M., "Pelleted Rations for Horses," *The Blood-Horse* (Lexington: 1964) p. 262.

Mohler, John, *et al.*, *Special Report on Diseases of the Horse* (Washington: U. S. Government Printing Office, 1942).

Patten, John, *The Light Horse Breeds* (New York: A. S. Barnes & Company, Inc., 1960).

Podhajski, Colonel A., *The Spanish Riding School of Vienna* (Vienna: Brüder Rosenbaum, 1958).

Robertson, William, ed., "Annual Statistical Review," *The Thoroughbred Record* (Lexington: 1964).

Rooney, James R., D.V.M., "Mare May Begin Preparations Long Before Actual Foaling," *The Blood-Horse* (1963) p. 998.

———, "Cleanliness is Best Safeguard Against Mycotic Placentitis," *The Blood-Horse* (Lexington: 1964).

Scheinfeld, Aram, *The New You and Heredity* (New York: J. B. Lippincott Company, 1950).

Schwender, Warren and Dan M. Bowmar III, "Feeding Practices on 12 Thoroughbred Farms," *The Blood Horse* (Lexington: 1956).

Simpson, George Gaylord, *Horses* (New York: Oxford University Press, 1951).

Sirron, Septimus, S.B., V.S., D.V.Sc., *et al, The Anatomy of Domestic Animals* (Philadelphia: W. B. Saunders Company, 1938).

Steele, Dewey G., "The Genetic Consideration Concerning White Foals," *The Blood-Horse* (Lexington: 1964) p. 225.

Stoneridge, M. A., *A Horse of Your Own* (New York: Doubleday & Company, Inc., 1963).

Teighland, M. B., D.V.M., "Sore Horses," *The Blood-Horse* (Lexington: 1961) p. 1471.

Tibbals, Lorren, "Gene is Million Dollar Daddy," *Akron Beacon Journal* (Akron: 1964).

Vickers, Captain Reginald J., M.C., *A Plea for Sanity in Thoroughbred Racing* (Richmond, Virginia: Whittet and Shepperson, 1962).

Wade, Horace, "Eclipsing Time," *The Thoroughbred Record* (Lexington: 1964) p. 1616.

Wentworth, Lady, *Thoroughbred Racing Stock* (London: George Allen & Unwin, Ltd., 1938).

Wright, Giles E., "Pelleted Feed Cuts Labor Costs at Laguna Seca," *The Blood-Horse* (Lexington: 1964) p. 156.

———, "Success Rewards Efforts at Laguna Seca," *The Blood-Horse* (Lexington: 1964) p. 82.

Index